Ghulam Shabir

HPLC Method Development and Validation in Pharmaceutical Analysis

D1158533

Ghulam Shabir

HPLC Method Development and Validation in Pharmaceutical Analysis

Handbook for Analytical Scientists

LAP LAMBERT Academic Publishing

Impressum / Imprint

Bibliografische Information der Deutschen Nationalbibliothek: Die Deutsche Nationalbibliothek verzeichnet diese Publikation in der Deutschen Nationalbibliografie; detaillierte bibliografische Daten sind im Internet über http://dnb.d-nb.de abrufbar.

Alle in diesem Buch genannten Marken und Produktnamen unterliegen warenzeichen-, marken- oder patentrechtlichem Schutz bzw. sind Warenzeichen oder eingetragene Warenzeichen der jeweiligen Inhaber. Die Wiedergabe von Marken, Produktnamen, Gebrauchsnamen, Handelsnamen, Warenbezeichnungen u.s.w. in diesem Werk berechtigt auch ohne besondere Kennzeichnung nicht zu der Annahme, dass solche Namen im Sinne der Warenzeichen- und Markenschutzgesetzgebung als frei zu betrachten wären und daher von jedermann benutzt werden dürften.

Bibliographic information published by the Deutsche Nationalbibliothek: The Deutsche Nationalbibliothek lists this publication in the Deutsche Nationalbibliografie; detailed bibliographic data are available in the Internet at http://dnb.d-nb.de.

Any brand names and product names mentioned in this book are subject to trademark, brand or patent protection and are trademarks or registered trademarks of their respective holders. The use of brand names, product names, common names, trade names, product descriptions etc. even without a particular marking in this works is in no way to be construed to mean that such names may be regarded as unrestricted in respect of trademark and brand protection legislation and could thus be used by anyone.

Coverbild / Cover image: www.ingimage.com

Verlag / Publisher:
LAP LAMBERT Academic Publishing
ist ein Imprint der / is a trademark of
AV Akademikerverlag GmbH & Co. KG
Heinrich-Böcking-Str. 6-8, 66121 Saarbrücken, Deutschland / Germany
Email: info@lap-publishing.com

Herstellung: siehe letzte Seite /
Printed at: see last page
ISBN: 978-3-659-32120-7

Handbook of Chromatographic Method Development and Validation in Pharmaceutical Analysis

Ghulam Shabir

Handbook of Chromatographic Method Development and Validation in Pharmaceutical Analysis

Ghulam Shabir BSc MSc PhD CSci CChem FRSC FCQI

Founder and Managing Director of DGS PharmaTraining Ltd, Oxford, United Kingdom

Research Fellow of Oxford Brookes University, United Kingdom

AUTHOR

Dr. Ghulam Shabir is Founder and Managing Director of DGS PharmaTraining Ltd, Oxford, United Kingdom. He is also a Research Fellow of OXFORD Brookes University.

Dr. Shabir has more than 27 years of leadership and management experience in the major international regulatory and GMP standards including World Health Organisation (WHO), Medicines and Healtcare products Regulatory Agency (MHRA) and U.S. Food and Drug Administration (FDA) compliance pharmaceutical industries and many academic institutions. He has held various positions as Analytical Scientist, Senior Scientist, Principal Scientist, Analytical Development Manager, Quality Control Manager, Quality Assurance Manager, Head of Quality Operations, visiting Professors, and Head of training with multinational organisations including Abbott Laboratories and several academic institutions, UK. During his career, he has received many industrial technical excellence and academic awards.

He received his graduate and post-graduate degrees in Chemistry from University of Sindh, post-graduate degree in Pharmaceutical Analysis from University of Strathclyde, UK, Higher Management studied at Glasgow Caledonian University, BTEC in Analytical Chemistry from University of Greenwich London, Managing Research & Development Projects professional advancement studied from University of Oxford, Communicative skills in English studied at University of Cambridge and DOCTORATE (PhD) in Analytical Chemistry obtained from the University of Sunderland.

Dr. Shabir is a Fellow of the Royal Society of Chemistry (FRSC), Chartered Chemist of the Royal Society of Chemistry (CChem), Chartered Scientist (CSci) of the Science Council UK, Fellow and Chartered Quality Professional (FCQI CQP) of the Chartered Quality Institute, London and Full Member of the American Society for Quality.

Dr. Shabir is the author of over 65 peer-reviewed papers, the author of 3 books, presented many posters, papers, lectures at international conferences and universities around the world. He is also a member of editorial boards for many high-impact journals and scientific magazines including American Pharmaceutical Review.

He has received Top Cited Article 2002 - 2007 Author Award for Journal of Chromatography A from Elsevier (Amsterdam) and two Top Cited Article 2010 Author Awards from Taylor & Francis (Philadelphia, USA) in recognition of his sustained exemplary service to the Journal of Liquid Chromatography & Related Technologies.

Dr. Shabir has won the 1998 Education Award from the Institute of Manufacturing UK. The competition was held throughout Europe and was awarded to the best article "Higher Management Role and Quality Assurance" written for the International Journal of Manufacturing.

Table of Contents

ACKNOWLEDGEMENTS

I would like to thank Dr. W John Lough (School of Health, Natural and Social Sciences, University of Sunderland), Dr. Tony K. Bradshaw (Department of Biological and Medical Sciences, Faculty of Health and Life Sciences, Oxford Brookes University) and Abbott Laboratories UK for their support and advice throughout this work.

None of this would have been possible without the love and support of my family, especially my parents for their encouragement. I also wish to express my sincere gratitude to my loving wife for her patience during this work and my sons Eyssa M Shabir and Zaid A Shabir, for their unconditional love and playfulness.

PREFACE

The coherent body of research described in this handbook is concerned with new chromatographic test method development and validation using novel systematic approaches for veriety of pharmaceutical and diagnostic compounds in a GMP environment.

The first stage of the research was to study how analytical method development and validation are typically carried out at present and to formulate this into a simple step-by-step approach. Such a template and protocol was not only used as the foundation of this research programme but could also serve as a simple systematic guide for other practitioners and those new to the field. Furthermore, it was recognised that this protocol should satisfy the requirements of the most strategically important international regulatory agencies. The second stage of this research involved evaluation and application of the above validation approach to new methods that were developed for a diverse range of analytes and samples.

A new purity assay for 1,10-phenanthroline-5,6-dione, 4,5-diazafluoren-9-one and benzo[f]quinoline-5,6-dione using high-performance liquid chromatography (HPLC) was developed and validated. Impurities in these compounds were identified by liquid chromatography-mass spectrometry (LC-MS). Best practice in method development and validation is equally important in the analysis of both active components and excipients in formulated products. In the first case, a liquid chromatography assay method for determining the content of nonoxynol-9; p+regn-4-ene-3,20-dione, 2-(diethylamino)-N-(2,6-dimethylphenyl) acetamide in a gel formulation; domperidone, sorbic acid and propylparaben in pharmaceutical formulations; phenothrin, methyl-4-hydroxybenzoate and propyl-4-hydroxybenzoate in human head lice medicine was developed and validated.

In the second case, methyl salicylate in a medicated cream formulation and sodium benzoate in liqu cough medicine by HPLC was developed and validated. Also the individual contents of seven p-hydroxybenzoic acid preservatives in a complex multi-component sample were determined following the development and validation of a HPLC method. In the third case, a HPLC assay method for determining the content of procainamide, n-acetylprocainamide, caffeine, methamphetamine and propranolol in tablet solid dosage formulation was developed and validated. In the fourth case, a liquid chromatography assay method of hydrolysed gelatine, 9α-fluoro-16β-methyl-prednisolone-17-valerate was developed and validated. In the fifth case, HPLC method for the determination of 11β,17α, 21-trihydroxypregn-4-ene-3,20-dione residues on manufacturing equipment surfaces was developed and validated.

Also the validation approach was evaluated as applied to another complex compounds. Here, HPLC novel assay method for nicotinamide adenine dinucleotide (NAD) in buffered solutions and dry test diagnostic strips presented different analytical problems because of the very complex nature of this natural product. Stability study information to increase the shelf life of the product and validation data for the analytical method for nicotinamide adenine dinucleotide content was critically evaluated. Also HPLC method for the determination of a series of eight barbiturates, phenothrin, methyl-4-hydroxybenzoate and propyl-4-hydroxybenzoate in human head lice medicine and determination of related substances of codeine phosphate in tablet formulation were developed and validated successfully.

Finally, the validation approach was evaluated as applied to another analytical technique. Here, gas chromatography (GC) successfully used to develop a novel assay for *p*-cymene in tea tree oil formulations presented different analytical problems because of the very complex nature of this natural product. Stability study information to increase the shelf life of the product and validation data for the analytical method for *p*-cymene content was critically evaluated.

In essence, the critical review of the requirements for method validation for various regulatory agencies and the subsequent preparation of guidelines on how to go about method validation have had a significant impact on how analytical practitioners worldwide go about method development and, more importantly, method validation. Further it was possible to apply these guidelines to conduct a series of effective, successful method validation for assays involving a range of typical pharmaceutical samples.

Ghulam Shabir, PhD
Oxford, UK

1 INTRODUCTION

Analytical method development and validation is an important part of analytical chemistry and plays a major role in the discovery, development, and manufacture of pharmaceuticals. The official test methods that result from these processes are used by quality control laboratories to ensure the identity, purity, potency and performance of drug product 'quality' essential for drug safety and efficacy.

In the pharmaceutical and biotechnology industries and even more so in the diagnostics industry, a current major issue is the high cost of research in introduction of new drugs. In essence it takes several hundred million dollars to discover, develop and gain regulatory approval. One of the reasons research and development (R&D) is so costly in pharmaceuticals is that most new drug candidates fail to reach the market. Failure can result from toxicity, carcinogenicity, manufacturing difficulties, inadequate efficacy and analytical problems. Therefore there is a need for high throughput in order to maximise patent lifetime and consequently generate the profits to support the research and to increase the speed with which products can be delivered to the market. All the different stages of pharmaceutical R&D are underpinned by analysis so that high throughput is acutely dependent on effective and efficient analysis within which simple effective method development and comprehensive analytical method validation is of fundamental importance.

A wide variety of materials are used in the pharmaceutical and diagnostic industries. All of these materials must be analysed in some way or other and, just as importantly, the method of analysis must be validated i.e. it must be shown that the method is fit for its intended purpose.

In the pharmaceutical field, method validation is very much a major issue as analysis is used primarily to control drug quality. This is important in its own right and also in that drug safety and efficacy are dependent on it. Different chemical entities with varying chemical and physical properties are used. These may include starting materials, intermediates, final drug substances and the final formulated pharmaceutical products. The pharmaceutical analyst will be concerned with applying analytical methods to the determination of stability/shelf life, purity, side-product identity, dissolution etc. Here, the analyst is required to develop new methods of analysis appropriate to the information required. In many cases, the analyte may be known but is present in a new sample matrix such that a new sample preparation method is needed. The knowledge gained in the method R&D phase is important when it comes to validating the research data efficiently. This is an important coherent theme that runs throughout the publications on which this thesis is based.

Frequently, high-performance liquid chromatography (HPLC) is the analytical method of choice in pharmaceutical analysis. Although HPLC is a relatively mature technique, the analyst is continually required to innovate by adapting current methodology or indeed developing completely new protocols. For example, the coupling of HPLC with another technique such as mass spectrometry (MS) can be an especially powerful tool.

In the diagnostic field, the variety of materials is further expanded due to the complexity of medical devices and their corresponding reagents. Such materials may include polymers, surfactants, enzymes, cofactors, mediators, stabilisers etc. The diagnostic analyst is therefore required to apply other techniques apart from HPLC in the analysis of key materials. An in-depth knowledge of the materials and their critical properties as applied to their use in the diagnostic device is necessary. Innovation is again needed if there is no directly applicable methodology reported in the literature. Once an analytical method is developed, validation is conducted in order to prove its results of the research are valid and its use for the intended application.

Method validation is a critical step for any product release for marketing authorization and is a regulatory requirement. The literature contains diverse approaches to performing method validation [1-12]. Many analytical methods appearing in the literature have not been through a thorough validation exercise and thus should be treated with caution until full validation has been carried out. Also there is no method validation reported with good manufacturing practice (GMP) considerations. Validation of a new method is a costly and very time-consuming exercise. However the result of not carrying out method validation could result in litigation, failure to get product approval, costly repeat analysis and loss of business and market share [13].

Currently, there is no completely worldwide single source or final guideline on method validation that helps analysts to perform validation in a systematic manner and most importantly under GMP considerations. Therefore industry depends on the analyst's knowledge and experience to develop simple and efficient methods of analysis.

The other major problem pharmaceutical industries are facing in today's world is that different validation data requirements are required for regulatory submissions for medicinal products' registration/approval depending upon the location of the regulatory body. For example the release of any medicinal product in USA, Europe and Japan would require the use of International Conference on Harmonisation (ICH) method validation criteria published in Q2(R1) "Validation of Analytical Procedures: Text and Methodology". However, the release of the very same product, by the same industry, in any other part of the world would force the use of their local regulatory guidelines. This inevitably becomes a costly process due to issues of documentation and personnel training etc. Therefore, efforts are underway to streamline the method validation process through an idea commonly referred to as Harmonisation by ICH. ICH is a joint initiative involving both regulators and industry as equal partners in the scientific and technical discussions of the testing procedures which are required to ensure and assess the safety, quality and efficacy of medicines. The focus of ICH has been on the technical requirements for medicinal products containing new drugs. The vast majority of those new drugs and medicines are developed in Western Europe, Japan and the United States of America and therefore, when ICH was established, it was agreed that its scope would be confined to registration in those three regions. Method validation is the proof needed to ensure that an analytical method can produce results that are valid, reliable, reproducible and are fit for the purpose intended. Choosing the validation criteria depends on the method type [14]. In general, method validation parameters that should be studied are linearity, range, accuracy, precision (repeatability and intermediate precision), specificity, limit of detection and limit of quantitation. Robustness and stability parameters should be considered in the method development phase. The detailed explanation of these parameters is given in chapter 2 and published in ICH Q2(R1) [15]. A step-by-step approach for method validation is published in author's paper [16].

The outcome of ICH efforts has been accepted by most regulatory agencies and pharmacopoeias such as U.S. Food and Drug Administration (FDA) and United States Pharmacopeia (USP) [17]. The ICH guidelines achieved a great deal in harmonising the definitions of the required validation characteristics and their basic requirements. However, they provide only a basis for a general discussion of the validation parameters, their calculation and interpretation. However, this has not removed the confusion in industries because ICH, as yet, has not explained various other method types such as response test (to detect a specific substance in a sample as indicated by test signal response), concentration test (for quantitation of a specific substance in a sample), physical test (for determination of the physical characteristics of a product or material) and cleaning test (for evaluating the cleanliness of equipment and areas used for manufacturing). Also ICH has not explained step-by-step approaches and most importantly GMP considerations that required during method validation to meet regulatory requirements.

The strategy described in this book is to address some specific objectives i.e. (i) to critically evaluate current practices in method development and validation in order to identify best practices as well as identify the key similarities and, more importantly, differences between the validation requirements of the FDA, USP and ICH. (ii) to apply best practices with some improvements, in such a way as to ensure good quality and provide new knowledge on a wide range of pharmaceutical substances, products and compounds used in pharmaceuticals, diagnostics and, finally, (iii) to draw upon the outcomes of the programme to be able to recommend the way forward with respect to ensuring that the ever-evolving approaches to analytical method development and validation were enhanced, simple, systematic, efficient and effective while still being compliant with the requirements of regulatory agencies. The second stage of the overall research programme involved evaluation and application of the above validation approaches to methods that were developed for a diverse range of analytes and complex samples using analytical techniques such as HPLC and LC-MS. These validated new assay methods have been implemented in GMP compliance quality control laboratories for routine analysis of raw materials, bulks, intermediates and final products release.

REFERENCES

1. Crowther, J.B. (2001). Validation of pharmaceutical test methods. *In: Handbook of Modern Pharmaceutical Analysis*, pp.415-543. California: Academic Press.
2. Wilson, T.D. (1990). Liquid chromatographic methods validation for pharmaceutical products.*J. Pharm. Biomed. Anal.* 8, 389-400.
3. Clarke, G.S. (1994). The validation of analytical methods for drug substances and drug products in UK pharmaceutical laboratories. *J. Pharm. Biomed. Anal.* 12, 643-652.
4. Bressolle, F., Petit, M.B. & Audran, M. (1996). Validation of liquid chromatographic and gas chromatographic methods: application to pharmacokinetics. *J. Chromatogr. B.* 686, 3-10.
5. Carr, G.P. & Wahlich, J.C. (1990). A practical approach to method validation in pharmaceutical analysis. *J. Pharm. Biomed. Anal.* 8, 613-618.
6. Green, M.J. (1996). A practical guide to analytical method validation. *Anal. Chem.* 68, 305A-309A.
7. Trullols, E., Ruisanchez, I. & Rius, F.X. (2004). Validation of qualitative analytical methods. *Trends Anal. Chem.* 23, 137-145.
8. Ermer, J. (2001). Validation in pharmaceutical analysis. Part 1: An integrated approach. *J. Pharm. Biomed. Anal.* 24, 755-767.
9. Daraghmeh, N., Al-Omari, et al. (2001). Determinaton of sildenafil citrate and related substances in the commercial products and tablet dosage form using HPLC. *J. Pharm. Biomed. Anal.* 25, 483–492.
10. Shabir, G.A., Lough, J.W., Arain, S.A. & Bradshaw, T.K. (2007a). Evaluation and application of best practice in analytical method validation. *J. Liq. Chromatogr. Rel. Technol.* 30, 311-333.

11. Badea, D., Ciutaru, L., Lazar, D.N. & Tudose, A. (2004). Rapid HPLC method for the determination of paclitaxel in pharmaceutical forms without separation. *J. Pharm.Biomed. Anal.* 34, 501–507.

12. Mendez, A.S.L., Steppe, M. et al. (2003). Validation of HPLC and UV spectrophotometric methods for the determination of meropenem in pharmaceutical dosage form. *J. Pharm. Biomed. Anal.* 33, 947–954.

13. Harvey, I.M., Baker, R. & Woodget, B. (2002). Part 2, the analytical method. In: Series of tutorial lectures: Quality assurance in analytical sciences, http://new.filter.ac.uk/database/getinsight.php?id=47&seq=2,

14. Shabir, G.A. (2003). Validation of HPLC methods for pharmaceutical analysis: Understanding the differences and similarities between validation requirements of the U.S. FDA, USP and the ICH. *J. Chromatogr. A.* 987(1-2), 57-66.

15. ICH, Q2(R1), Validation of Analytical Procedures: Text and methodology, Geneva, November 2005.

16. Shabir, G.A. (2004). Step-by-step analytical methods validation and protocol in the quality system compliance industry, *J. Validation Technol.* 10(4), 314-324.

17. U.S. Pharmacopeia 34. (2011). General chapters (1225) Validation of compendial procedures, United States Pharmacopeal Convention, Inc., Rockville Maryland.

2 ADVANCES IN CHROMATOGRAPHIC TECHNIQUES FOR PHARMACEUTICAL ANALYSIS

This section is intended to give a short overview of chromatographic history and instrumentation, operating principles and recent advances or trends that are used in the pharmaceutical analysis.

1. History and Development of Chromatography

Liquid chromatography was defined in the early 1900s by the work of the Russian botanist, Mikhail S. Tswett. His pioneering studies focused on separating compounds (leaf pigments), extracted from plants using a solvent in a column packed with particles. Tswett filled an open glass column with particles. Two specific materials that he found useful were powdered chalk (calcium carbonate) and alumina. He poured his sample (solvent extract of homogenized plant leaves) into the column and allowed it to pass into the particle bed. This was followed by pure solvent. As the sample passed down through the column by gravity, different colored bands could be seen separating because some components were moving faster than others. He related these separated, different-colored bands to the different compounds that were originally contained in the sample. He had created an analytical separation of these compounds based on the differing strength of each compounds chemical attraction to the particles. The compounds that were more strongly attracted to the particles slowed down, while other compounds more strongly attracted to the solvent moved faster. This process can be described as follows: the compounds contained in the sample distribute, or partition differently between the moving solvent, called the mobile phase, and the particles, called the stationary phase. This causes each compound to move at a different speed, thus creating a separation of the compounds.Tswett coined the name *chromatography* (from the Greek words *chroma*, meaning color, and *graph*, meaning writing - literally, *color writing*) to describe his colorful experiment.

Since the various types of chromatography (liquid, gas, paper, thin-layer, ion exchange, supercritical fluid, and electrophoresis) have many feauters in common, they must all be considered in development of the field. Since the early 1970s, results of these studies have had a significant impact on the other types of chromatography, especially modern high-performance liquid chromatography (HPLC). Today, HPLC has become one of the most powerful tools in analytical chemistry.

2. High Performance Liquid Chromatography

The acronym HPLC, coined by the late Prof. Csaba Horváth for his 1970 Pittcon paper, originally indicated the fact that high pressure was used to generate the flow required for liquid chromatography in packed columns. In the beginning, pumps only had a pressure capability of 500 psi (35 bar). This was called *high pressure* liquid chromatography. The early 1970s saw a tremendous leap in technology. These new HPLC instruments could develop up to 6,000 psi (400 bar) of pressure, and incorporated improved injectors, detectors, and columns. HPLC really began to take hold in the mid-to late-1970s. With continued advances in performance during this time (smaller particles, even higher pressure), the acronym HPLC remained the same, but the name was changed to *high performance* liquid chromatography. HPLC is now one of the most powerful tools in analytical chemistry. It has the ability to separate, identify, and quantitate the compounds that are present in any sample that can be dissolved in a liquid.

Today, compounds in trace concentrations as low as *parts per trillion* (ppt) may easily be identified. HPLC can be, and has been, applied to just about any sample, such as pharmaceuticals, food, nutraceuticals, cosmetics, environmental matrices, forensic samples, and industrial chemicals. A modern HPLC system is shown systematically in Figure 1.

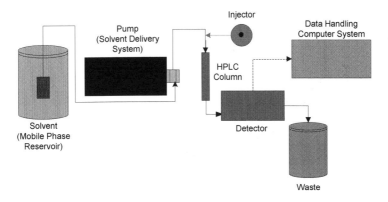

FIGURE 1 The flow diagram of general HPLC system.

The equipment consists of a high-pressure solvent delivery system, a sample auto injector, a separation column, a detector (often photodiode array (PDA) ultraviolet visible (UV-Vis) detector a computer to control the system and display results. Many systems include an oven for temperature control of the column and a pre-column that protects the analytical column from impurities. The actual separation takes place in the column, which is packed with chemically modified 3.5-10 μm (often silica) particles. A mobile phase is pumped through the column with the high-pressure pump and the analytes in the injected sample are separated depending on their degree of interaction with the particles. A proper choice of stationary and mobile phase is essential to reach a desired separation.

3   HPLC Separation Modes

In general, three primary characteristics of chemical compounds can be used to create HPLC separations. They are:
•Polarity
•Electrical Charge
• Molecular Size
First, let's consider polarity and the two primary separation modes that exploit this characteristic: normal phase and reversed-phase chromatography.

3.1 Separations Based on Polarity

A molecule's structure, activity, and physicochemical characteristics are determined by the arrangement of its constituent atoms and the bonds between them. Within a molecule, a specific arrangement of certain atoms that is responsible for special properties and predictable chemical reactions is called a functional group. This structure often determines whether the molecule is *polar* or *non-polar*. Organic molecules are sorted into classes according to the principal functional group(s) each contains. Using a separation mode based on polarity, the relative chromatographic retention of different kinds of molecules is largely determined by the nature and location of these functional groups.

To design a chromatographic separation system, we create competition for the various compounds contained in the sample by choosing a mobile phase and a stationary phase with different polarities. Then, compounds in the sample that are similar in polarity to the stationary phase (column packing material) will be delayed because they are more strongly attracted to the particles. Compounds whose polarity is similar to that of the mobile phase will be preferentially attracted to it and move faster. In this way, based upon differences in the relative attraction of each compound for each phase, a separation is created by changing the speeds of the analytes.

3.2 Normal-Phase HPLC

In normal-phase chromatography (NPC) the stationary phase is more polar than the mobile phase. Usually, the mobile phase is a mixture of organic solvents without added water (e.g., isopropanol, hexane) and the column packing is either an inorganic adsorbent (silica or occasionally alumina) or a polar bonded phase (cyano, diol, or amino) on a silica support. Regardless of the mobile phase or stationary phase used, sample retention in NPC increases as the polarity of the mobile phase decreases (the opposite of the reverse-phase chromatography).

3.3 Reversed-Phase HPLC

The term reversed-phase describes the chromatography mode that is just the opposite of normal phase, namely the use of a polar mobile phase and a non-polar (hydrophobic) stationary phase.

Today, reversed-phase chromatography is probably the most commonly used separation mechanism in liquid chromatography, because it is more reproducible and has broad applicability for all HPLC methods. It consists of a non-polar (hydrophobic) stationary phase (normally octadecyl, C_{18} (sometimes called ODS) or octyl C_8 chains) bonded to a solid support that is generally micro particulate silica gel. Silica has a small pH range (3-8) where mixtures can be separated without degradation of the column performance. The mobile phase is polar and, therefore, the sample compounds are partitioned between the mobile and the stationary phases. The separation is normally performed using aqueous mobile phase containing different percentages of organic modifiers (e.g. methanol, ethanol, acetonitrile, or tetrahydrofuran) to increase the selectivity between species. Solute retention is also influenced by eluent pH, which affects the dissociation level of the analyte and therefore, its partition between the mobile and stationary phases. Table 1 presents a summary of the phase characteristics for the two principal HPLC separation modes based upon polarity. Remember, for these polarity-based modes, *like attracts like*.

TABLE 1 Phase Characteristics for Separations Based on Polarity

Separation Mode	Stationary Phase (Particle)	Mobile Phase (Solvent)
Normal Phase	Polar	Non-polar
Reversed phase	Non-polar	Polar

4. Ultra-Performance Liquid Chromatography (UPLC)

In 2004, further advances in instrumentation and column technology were made to achieve very significant increases in resolution, speed, and sensitivity in liquid chromatography. Columns with smaller particles (1.7 micron) and instrumentation with specialized capabilities designed to deliver mobile phase at 15,000 psi (1,000 bar) were needed to achieve a new level of performance. A new system had to be holistically created to perform ultra-performance liquid chromatography, now known as UPLC technology.

Basic research is being conducted today by scientists working with columns containing even smaller 1-micron-diameter particles and instrumentation capable of performing at 100,000 psi (6,800 bar). This provides a glimpse of what we may expect in the future.

5. UV-VIS Absorbance and Photodiode-Array Detection

The UV-Vis detectors are the most common HPLC detectors and frequently used to measure components showing an absorption spectrum in the ultraviolet or visible region. A UV detector employs a deuterium discharge lamp (D2 lamp) as a light source, with the wavelength of its light ranging from 190 to 380 nm (UV range). If components are to be detected at wavelength longer than this, a UV-VIS detector is used, which employs an additional tungsten lamp (Figure 2, W lamp). Figure 2 shows the optical system. Light from the lamp is shone onto the diffraction grating, and dispersed according to wavelength. For example, when the measurement is performed with a wavelength of 280 nm, the angle of the diffraction grating is adjusted so that 280 nm light can shine on the flow cell. By monitoring the reference light divided from the light in front of the flow cell, the difference in light intensity can be determined between the back and front of the flow cell, and this is output as absorbance.

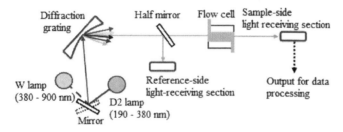

FIGURE 2 Schematic of a UV-Vis absorbance detector optical system.

The ordinary UV detector measures the absorbance at one single wavelength at the time. To change wavelength the monochromator must be moved, therby introducing the problems of mechanical irreproducibility into the measurements. A photodiode array (PDA) detector (Figure 3) can measure multiple wavelengths at the same time, and since no parts are moved to change wavelength or to scan, there are no mechanical errors or drift with time. Following one chromatographic run with a PDA enables the analyst to display a chromatogram for any desired wavelength and in addition the UV spectra of each eluting peak can be displayed. PDAs therefore provide more information on sample composition than provided by a single wavelength run. While single wavelengths detector are commonly used for quantitative analysis, a PDA can be used for both quantitative and qualitative information of the samples. If the peaks in chromatograms are resolved, UV spectra can be collected for each peak with help of a PDA, and peak identification during method development and peak purity evaluation during method validation can be carried out with the recorded signals. Peak identification is made by comparing the UV spectra for sample peak with a reference standard peak by overlapping the two spectra. Even if a standard is not available it can reveal if a components represented by a peak in a chromatogram is related to another compound.

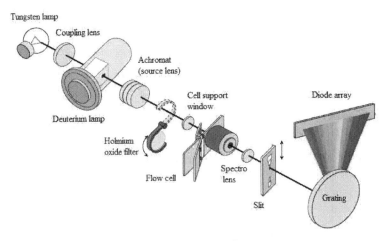

FIGURE 3 Schematic of a photodiode array detector.

6. Liquid Cromatography-Mass Spectrometry

Mass spectrometry (MS) has progressed extremely rapidly during the last decade: production, separation and detection of ions, data acquisition, data reduction, etc. and this has led to the development of entirely new modern instruments and applications.

The combination of chromatographic separations with mass spectrometric detection is considered an indispensable tool for problem solving in analytical chemistry and increasingly for routine analytical methods. Mass spectrometric detection brings an added level of information, complementary to the chromatographic process that improves the certainty of identification and the specificity of detection. Mass spectral information can generally be obtained from sample sizes typical of common analytical methods. In the last ten years, research efforts in the field of LC-MS have changed considerably.

LC-MS (Figure 4) has rapidly matured to become a very powerful and useful analytical tool that is widely applied in many areas of chemistry, pharmaceutical sciences and biochemistry. Investigation into the coupling of HPLC and MS began in the early 1970s.

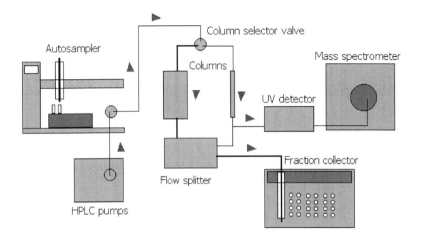

FIGURE 4 LC-MS system.

In the first 20 years, most of the attention had to be given to solving interface problems and building new technology. However, most scientists with LC-MS today are only concerned with application of the commercially available techniques in their field of interest. Technological problems in interfacing appear to be solved, and from the wide variety of interfaces developed over the years basically only two dominate, i.e. electrospray ionisation (ESI) and atmospheric-pressure chemical ionisation (APCI) (Figure 5), which are both atmospheric-pressure ionisation (API) techniques.

With LC-MS, ESI and APCI has been implemented in analytical strategies in many application areas, e.g. environmental analysis, drug development within the pharmaceutical industry, characterisation of natural products and the characterisation of biomolecules like peptides, proteins, oligosaccharides etc.

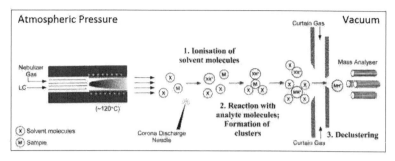

FIGURE 5 Atmospheric pressure chemical ionization process.

The selection of the appropriate HPLC conditions, whether reversed-phase liquid chromatography, ion-pairing chromatography, capillary electrophoresis or ion chromatography, and ionisation mode, ESI or APCI, depends upon the polarity of the analyte. ESI is best applied to the highly polar nature of the analyte and APCI ionises most efficiently compounds with low to moderately high polarities and in this respect is complementary to electrospray, which gives the best sensitivity for ionic compounds. Both interfaces ESI and APCI can be operated in positive and negative ion mode. Often, an appropriate selection for a given analyte can be made by considering that ESI transfers ions from solution into the gas phase, whereas APCI ionizes in the gas phase. As a rule of thumb, analytes occurring, as ions in solution may be best analysed by ESI, while non-ionic analytes may be well suited for APCI. Figure 6 shows the principle of the ESI process in positive ion mode.

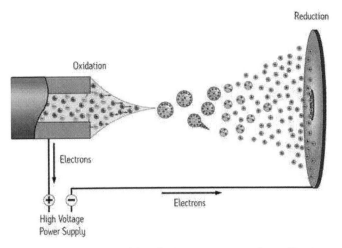

FIGURE 6 Illustration of the electrospray process in positive ion mode. (Reprinted with permission from Andreas Dahlin)

As for the detection principles discussed above, all of these contribute significantly to the present-day success of hyphenation in HPLC. There is no doubt that, also today HPLC-PDA/UV-Vis detector plays an important role (detection and peak-purity) in many research and development studies and for a wide variety of routine analyses in pharmaceutical industry.

7. Gas Chromatography

Gas chromatography (GC) is involves a sample being vapourised and injected onto the head of the chromatographic column. The sample is transported through the column by the flow of inert, gaseous mobile phase. The column itself contains a liquid stationary phase which is adsorbed onto the surface of an inert solid. Schematic diagram of a gas chromatograph is given in Figure 7.

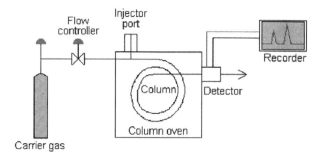

FIGURE 7 Schematic diagram of a gas chromatograph.

The carrier gas must be chemically inert. Commonly used gases include nitrogen, helium, argon, and carbon dioxide. The choice of carrier gas is often dependant upon the type of detector which is used. The carrier gas system also contains a molecular sieve to remove water and other impurities.

For optimum column efficiency, the sample should not be too large, and should be introduced onto the column as a "plug" of vapour - slow injection of large samples causes band broadening and loss of resolution. The most common injection method is where a microsyringe is used to inject sample through a rubber septum into a flash vapouriser port at the head of the column. The temperature of the sample port is usually about 50°C higher than the boiling point of the least volatile component of the sample. For packed columns, sample size ranges from tenths of a microliter up to 20 microliters. Capillary columns, on the other hand, need much less sample, typically around $10^{-3}/L$. For capillary GC, split/splitless injection is used (Figure 8).

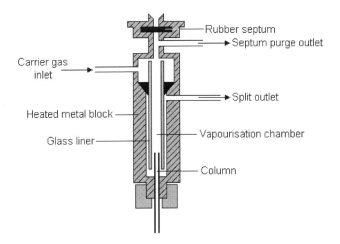

Carrier gas inlet

Heated metal block

Glass liner

Rubber septum

Septum purge outlet

Split outlet

Vapourisation chamber

Column

FIGURE 8 The Split/splitless injector.

The injector can be used in one of two modes; split or splitless. The injector contains a heated chamber containing a glass liner into which the sample is injected through the septum. The carrier gas enters the chamber and can leave by three routes (when the injector is in split mode). The sample vapourises to form a mixture of carrier gas, vapourised solvent and vapourised solutes. A proportion of this mixture passes onto the column, but most exits through the split outlet. The septum purge outlet prevents septum bleed components from entering the column.

There are two general types of column, *packed* and *capillary* (also known as *open tubular*). Packed columns contain a finely divided, inert, solid support material (commonly based on *diatomaceous earth*) coated with liquid stationary phase. Most packed columns are 1.5 - 10m in length and have an internal diameter of 2 - 4mm.

Capillary columns have an internal diameter of a few tenths of a millimeter. They can be one of two types; *wall-coated open tubular* (WCOT) or *support-coated open tubular* (SCOT). Wall-coated columns consist of a capillary tube whose walls are coated with liquid stationary phase. In support-coated columns, the inner wall of the capillary is lined with a thin layer of support material such as diatomaceous earth, onto which the stationary phase has been adsorbed. SCOT columns are generally less efficient than WCOT columns. Both types of capillary column are more efficient than packed columns.

For precise work, column temperature must be controlled to within tenths of a degree. The optimum column temperature is dependant upon the boiling point of the sample. As a rule of thumb, a temperature slightly above the average boiling point of the sample results in an elution time of 2 - 30 minutes. Minimal temperatures give good resolution, but increase elution times. If a sample has a wide boiling range, then temperature programming can be useful. The column temperature is increased (either continuously or in steps) as separation proceeds.

There are many detectors which can be used in gas chromatography. Different detectors will give different types of selectivity. A *non-selective* detector responds to all compounds except the carrier gas, a *selective detector* responds to a range of compounds with a common physical or chemical property and a *specific detector* responds to a single chemical compound. Detectors can also be grouped into *concentration dependant detectors* and *mass flow dependant detectors*. The signal from a concentration dependant detector is related to the concentration of solute in the detector, and does not usually destroy the sample dilution of with make-up gas will lower the detectors response. Mass flow dependant detectors usually destroy the sample, and the signal is related to the rate at which solute molecules enter the detector. The response of a mass flow dependant detector is unaffected by make-up gas.

The flame-ionization (FID) detector is the most common detector on commercial gas chromatographs is shown in Figure 9, consisting essentially of a base in which the column eluent is mixed with hydrogen, a polarised jet and a cylindrical electrode arranged concentric with the flame. Air is supplied to the detector to support combustion. The assembly is contained in a stainless steel or aluminium body to which are fitted a flame ignition coil and electrical connections to the collecting electrode, and a polarising voltage to the detector jet.

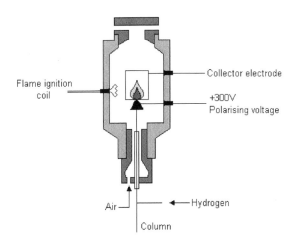

FIGURE 9 The flame ionisation detector.

The FID are mass sensitive rather than concentration sensitive, this gives the advantage that changes in mobile phase flow rate do not affect the detector's response. The FID is a useful general detector for the analysis of organic compounds; it has high sensitivity, a large linear response range, and low noise. It is also robust and easy to use.

REFERENCES

1. G. Shabir, Practical HPLC and LC-MS method Development and Validation, LAP Lambert Academic Publishing, Germany, 2012.
2. R. Grob, E.Barry, Modern Practice of Gas Chromatography, John Wiley & Sons, New Jersey, 2004.

3 VALIDATION OF HPLC METHODS FOR PHARMACEUTICAL ANALYSIS: UNDERSTANDING THE DIFFERENCES AND SIMILARITIES BETWEEN FDA, USP AND ICH GUIDELINES

ABSTRACT

One of the most critical factors in developing pharmaceutical drug substance and drug products today is ensuring that the HPLC analytical test methods that are used to analyse the products generate meaningful data. U.S. Food and Drug Administration (FDA) and United States Pharmacopeia (USP) have each recognised the importance of this to the drug development process and have separately increased validation requirements in recent years. A third source, the International Conference on Harmonization (ICH), has added requirements that, when combined with the previous two sources, has led to three different sets of validation requirements leaving the industry in a state of confusion. This paper is written to clear up the confusion over the validation requirements that are presented by each of these three sources.

Keywords: HPLC, Analytical method validation, CGMP, FDA, USP, ICH

1. INTRODUCTION

Analytical method validation is completed to ensure that an analytical methodology is accurate, specific, reproducible and robust over the specified range that an analyte will be analyzed. Method validation provides an assurance of reliability during normal use, and is sometime referred to as "the process of providing documented evidence that the method does what it is intended to do." Regulated laboratories must perform method validation in order to be in compliance with FDA regulations. For pharmaceutical high performance liquid chromatography (HPLC) methods validation, guidelines from the FDA [1, 2], USP [3] and ICH [4] provides a framework for performing such validation (see Table 2). Method validation has received considerable attention in the literature [5-11] and regulatory agencies. The FDA has proposed in addition to adding section 211.222 on method validation to the current Good Manufacturing Practice (cGMP) regulations [5]. This would require the manufacturer to establish and document the accuracy, sensitivity, specificity, reproducibility and any other attribute necessary to validate test methods. Unfortunately, there is no single source or final guideline on analytical methods validation. Validation is customized by choosing necessary tests and acceptance criteria for a given method. The comprehensiveness of this kind of validation is based upon the type of method and its requirements. This article begins with a discussion of the overall process of analytical test method validation, including with instrument qualification as pre-validation requirement. Then, the subject of validation is addressed on the basis of currently accepted FDA, USP terminology and methodology, incorporating a discussion of the new ICH guidelines.

2. STEP-BY-STEP TO ESTABLISH METHOD VALIDATION PLAN

The first step in the development of a method validation protocol is to determine the objective of the method. How will the method be used? What is the method intended to demonstrate? Based on the response to these questions, there will be at least two main choices. For example, if the method is intended to release final product, or determine potency, level of impurity, or contaminants in a human drug product, the method is considered a level 1 (quantitative assay). If the method is intended to serve as a qualitative evaluation for identity, the method is considered a Level II. Is the method to be used for establishing a limit of impurity (less than or greater than a standard)? Are the results visual? In all cases, the following additional questions will also need to be answered: What sample-types will be tested using the method? Will the samples be whole blood, serum, plasma, purified protein, unpurified protein, chemical agents, etc.? Based on the sample-type, what interferences are expected? Is it likely that those interfering substances will impact the results? Is the method cell-based, chemical-based or enzyme-based? What level of accuracy, precision, sensitivity and limit of detection is required? The goal of the questions and the preliminary evaluation is to determine how best to meet the objective of the method validation so that it can be documented as suitable for its intended use.

The preparation and execution should follow a validation standard operating procedure (SOP), preferably written in a clear step-by-step instructions format. These include the following possible steps in analytical test method validation: assemble a cross-functional team and assign to individual responsibilities; define the purpose and scope of the method; determine the validation approach, method type and corresponding analytical performance characteristics; prepare a validation SOP; set the acceptance criteria on the basis of method development/characterisation data; write the test method as provisional use only format; perform pre-validation experiments; adjust method parameters and/or acceptance criteria if necessary; approve the validation SOP; execute the validation SOP and evaluate the results; prepare the validation report, review and approve.

2.1 Assemble a Validation Cross-Functional Team

Members of the cross functional team, assembled by the method validation project controller include representatives from the following departments including analytical development, quality control (end user laboratory management), quality assurance and the individuals requiring the analytical data. The validation SOP and the validation master plan (VMP) and or validation protocol should clearly define the roles and responsibilities of each individual involved in the method validation project, e.g., who will prepare the validation protocol, report and review and approve validation documents. All individuals assigned to the validation project should be adequately trained with respect to safety when handling chemicals, biological agents, etc. They should also be trained on the use of the equipment and protocol. Training records should be maintained and competency should be assessed.

3. PRE-VALIDATION REQUIREMENTS

In all types of analytical method validation, all required pre-qualification must be performed. The following items must be evaluated and more extensive evaluation is necessary for those that may have a higher potential to affect the assay.

3.1 Analytical Equipment Validation

Before to undertaking the task of method validation, it is necessary that the analytical system itself is adequately designed, maintained, calibrated and tested. In all cases proper validation documentation should be archived to support the qualification process. As can be seen in Figure 1, validation begins at the vendor's site, in a structural validation stage. During this stage, the analytical instrument and software are developed, designed and produced in a validated environment according to good laboratory practices (GLP), cGMP, and / or International Organization for Standardization (ISO) and others, for example ISO/IEC 17025 [12].

Recently, U.S. FDA has published a guideline 21 CFR Part 11 [13] focuses on software validation of computer systems. During the functional validation or qualification stage, the Installation Qualification (IQ), Operational Qualification (OQ), and Performance Qualification (PQ) are performed. The IQ establishes that the instrument is received as designed and specified and that it is properly installed. The OQ process ensures that the specific modules of the system are operating according to the defined specifications for accuracy, linearity and precision. This process may be as simple as verifying the module self diagnostic routines, or may be performed in more depth by running specific tests to verify, for example, detector wavelength accuracy, flow rate or injector precision. The PQ step verifies system performance. PQ testing is conducted under actual running conditions across the anticipated working range.

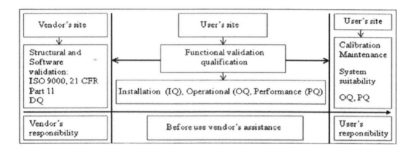

FIGURE 1 Analytical equipment qualification timeline that required before analytical test method validation.

For HPLC, the PQ test should use a method with a well-characterized analyte mixture. It should incorporate the essence of the system suitability section of the general chromatography section (<621>) in the USP [6]. After the instrument is placed on-line in the laboratory, and after a set period of use, regulations require maintenance followed by calibration and standardization, sometimes referred to as maintenance procedures. A system suitability test provides assurance that a system's performance still is appropriate for use.

3.2 Stability of Analytical Solutions

To generate reproducible and reliable result, the stability of sample solutions, standards, reagents and mobile phases must be determined prior to initiating the method validation studies. It is often essential that solutions be stable enough to allow for delays such as instrument breakdowns or overnight analyses using autosamplers. Samples and standards should be tested over at least a 24h period (depends on need), and quantitation of components should be determined by comparison to freshly prepared standards. A stability criterion for assay methods is that sample and standard solutions and the mobile phase will be stable for 24 h under defined storage conditions. Acceptable stability is 2% changes in standard or sample response, relative to freshly prepared standards. The mobile phase is considered to have acceptable stability if aged mobile phase produces equivalent chromatography (capacity factors, resolution, or tailing factor) and assay results are within 2% of the value obtained with fresh mobile phase. For impurity methods, the sample, standard solutions and mobile phase will be stable for 24h under defined storage conditions. Acceptable stability is 20% change in standard or sample response at the limit of quantitation, relative to freshly prepared standards. The mobile phase is considered to have acceptable stability if aged mobile phase produces equivalent chromatography and if impurity results at the limit of quantitation are within 20% of the values obtained with fresh mobile phase. If a solution is not stable at room temperature, then decreasing the temperature to 2-8°C can improve stability of samples and standards.

3.3 System Suitability Test

The quality of HPLC data collected begins with a well-behaved chromatographic system. Before performing any test method validation experiments, you should establish that the HPLC system and procedure are capable of providing data of acceptable quality. These tests are used to verify that the resolution and repeatability of the system are adequate for the analysis to be performed. System suitability tests are based on the concept that the equipment, electronics, analytical operations, and samples constitute an integral system that can be evaluated as a whole. System suitability is the checking of a system to ensure system performance before or during the analysis of unknowns. Parameters such as plate count, tailing factors, resolution and repeatability (%RSD retention time and area for six repetitions) are determined and compared against the specifications set for the method. The parameter to be measured and their recommended limits [1] obtained from the analysis of the system suitability sample are shown in Table 1. The quality control sample and standard are strongly recommended in the system suitability testing. The sample and standard should be dissolved in the mobile phase. If that is not possible, then avoid using too high a level of the organic solvent as compared to the level in the mobile phase. The concentration of sample and standard should be close if not the same and samples should be bracketed by standards during the HPLC analysis.

TABLE 1 System suitability parameters and recommendations

Parameter	Recommendation
Capacity factor (k')	The peak should be well-resolved from other peaks and the void volume, generally k'>2.0
Repeatability	RSD </= 1% for N >/= 5 is desirable.
Relative retention	Not essential as long as the resolution is stated.
Resolution (R_s)	R_s of > 2 between the peak of interest and the closest eluting potential interference (impurity, excipient, degradation product, internal standard, etc.
Tailing factor (T)	T of </= 2
Theoretical plates (N)	In general should be > 2000

The capacity factor (k') (occasionally called retention factor) is a measure of where the peak of interest is located with respect to the void volume, i.e., elution time of the non-retained components. The peak should be well resolved from other peaks and the void volume.

$$K' = (t_R - t_0) / t_0 \qquad \text{(Eq. 1)}$$

where t_0 is elution time of the void volume of non-retained components and t_R is the retention time of the analyte.

Resolution (R_s) is a measure of how well two peaks are separated. For reliable quantitation, well-separated peaks are essential for quantitation. This is a very useful parameter if potential interference peaks may be of concern. The potential peak eluting closest to the main analyte peak of interest should be selected. R_s is minimally influenced by the ratio of the two compounds being measured. R_s of > 2 between the peak of interest and the closest potentially interfering peak (impurity, excipient, degradation product, internal standard, etc.) is desirable.

$$R_s = \quad 2\,(t_{R2} - t_{R1})\,/\,(w_2 + w_1) \qquad \text{(Eq. 2)}$$

where t_w is peak width measured at baseline of the extrapolated straight sides to baseline.

Relative retention or selectivity (α): (also called selectivity or separation factor α value) is a measure of the relative location of two peaks. This is not an essential parameter as long as the resolution (R_s) is stated.

$$\alpha = k_2\,/\,k_1 \qquad \text{(Eq. 3)}$$

Tailing factor (T) is the accuracy of quantitation decreases with increase in peak tailing because of the difficulties encountered by the integrator in determining where/when the peak ends and hence the calculation of the area under the peak. Integrator variables are pre-set by the analyst for optimum calculation of the area for the peak of interest.

$$T = W_x\,/\,2f \qquad \text{(Eq. 4)}$$

where W_x is width of the peak determined at either 5% (0.05) or 10% (0.10) from the baseline of the peak height. f is distance between peak maximum and peak front at W_x

Theoretical plate number (N): N is a measure of column efficiency, that is, how many peaks can be located per unit run-time of the chromatogram.

$$N = 16 \, (t_R / t_w)^2 = L / H \qquad \text{(Eq. 5)}$$

N is fairly constant for each peak on a chromatogram with a fixed set of operating conditions. H, or the height equivalent of a theoretical plate (HETP), measures the column efficiency per unit length (L) of the column. Parameters, which can affect N or H, include peak position in the chromatogram particle size of stationary phase packing, flow-rate of mobile phase, column temperature, viscosity of mobile phase, and molecular weight of the analyte. The theoretical plate number depends on the elution time but in general should be > 2000.

4. DIFFERENCES AND SIMILARITIES BETWEEN FDA, USP AND ICH

To continue the discussion of method validation, it is necessary to have a complete understanding of the terminology and definitions involved. ICH has published the guideline: Validation of Analytical Procedures: Text and Methodology. ICH divided the 'validation characteristics' somewhat differently to USP as outlined in Table 2.

The difference in the USP and ICH terminology is, for the most part, one of semantics – with one notable exception. ICH treats system suitability as a part of method validation, whereas the USP treats it in a separate section (<621>) chromatography [6]. As this guideline has reached step 4 of the ICH process, FDA has begun to implement it, and it is anticipated that the ICH text and methodology will eventually be published in the USP. What follows then is a discussion of current USP definitions of the analytical performance parameters compared and contrasted with the ICH definitions. Where appropriate, methodology is also presented according to the ICH guideline on this subject.

TABLE 2 The ICH, USP and FDA validation parameters requirements

ICH/USP validation parameter	Additional FDA validation requirement	FDA CGMP (legal) requirement	Other type of method [4]
Specificity [1]	Sensitivity	Accuracy	Cleaning
Accuracy [1]	Recovery	Sensitivity	Specific tests
Precision: Repeatability [1]	Reproducibility	Specificity	Concentration
Intermediate precision [1]	Robustness	Reproducibility	Response
Reproducibility [3]	Stability		Physical method
Detection and	System		
Quantitation limits [1]	suitability		
Linearity and Range [1]			
Ruggedness [2] and			
Robustness [2,3]			

[1] ICH and USP requirement; [2] Included in the USP; [3] Included in ICH publication but not part of required parameter; [4] Not included yet in ICH publication

4.1 Accuracy

Accuracy is the closeness of the test results obtained by the analytical method to the true value. Accuracy is usually determined in one of four ways. First, accuracy can be assessed by analyzing a sample of known concentration (reference materials) and comparing the measured value to the true value. The second approach is to compare test results from the new method with results from an existing alternate well-characterized procedure that is known to be accurate. The third approach is based on the recovery of known amounts of analyte is performed by spiking analyte in blank matrices. For assay methods, spiked samples are prepared in triplicate at three levels over a range of 50-150% of the target concentration. For impurity methods, spiked samples are prepared in triplicate at three levels over a range that covers the expected impurity content of the sample, such as 0.1-2.5 wt%. The analyte levels in the spiked samples should be determined using the same quantitation procedure as will be used in the final method procedure (i.e., same number and levels of standards, same number of sample and standard injections, etc.). The percent recovery should then be calculated. The fourth approach is the technique of standard additions, which can also be used to determine recovery of spiked analyte. This approach is used if it is not possible to prepare a blank sample matrix without the presence of the analyte. This can occur, for example, with lyophilized material, in which the speciation in the lyophilized material is significantly different when the analyte is absent. Accuracy criteria for an assay method (FDA) is that the mean recovery will be 100 ±2% at each concentration over the range of 80-120% of the target concentration. For an impurity method, the mean recovery will be within 0.1% absolute of the theoretical concentration or 10% relative, whichever is greater, for impurities in the range of 0.1-2.5 wt%. To document accuracy the ICH guideline on methodology recommends collecting data from a minimum of nine determinations over a minimum of three concentration levels covering the specified range (for example, three concentrations, three replicates each).

4.2 Precision

Precision is the measure of the degree of repeatability of an analytical method under normal operation and is normally expressed as the percent relative standard deviation for a statistically significant number of samples. According to the ICH [4], precision should be performed at three different levels: repeatability, intermediate precision, and reproducibility. Repeatability is the results of the method operating over a short time interval under the same conditions (intra-assay precision). It should be determined from a minimum of nine determinations covering the specified range of the procedure (for example, three levels, three repetitions each) or from a minimum of six determinations at 100% of the test or target concentration.   According to the FDA [1] for instrument precision or injection repeatability study, should be a minimum of 10 injections of one sample solution is made to test the performance of the chromatographic instrument. Intermediate precision is the results from within lab variations due to random events such as different days, analysts, equipment, etc. In determining intermediate precision, experimental design should be employed so that the effects (if any) of the individual variables can be monitored. Precision criteria for an assay method is that the instrument precision (RSD) will be ≤1% and the intra-assay precision will be ≤2%. For impurity assay, at the limit of quantitation, the instrument precision will be ≤5% and the intra-assay precision will be ≤10%. Reproducibility [4], which is determined by testing homogeneous samples in multiple laboratories, often as part of inter-laboratory crossover studies. The evaluation of reproducibility results often focuses more on measuring bias in results than on determining differences in precision alone. Statistical equivalence is often used as a measure of acceptable inter-laboratory results. An alternative, more practical approach is the use of "analytical equivalence" in which a range of acceptable results is chosen prior to the study and used to judge the acceptability of the results obtained from the different laboratories.

An example of reproducibility criteria for an assay method could be that the assay results obtained in multiple laboratories will be statistically equivalent or the mean results will be within 2% of the value obtained by the primary testing laboratory. For an impurity method, results obtained in multiple laboratories will be statistically equivalent or the mean results will be within 10% (relative) of the value obtained by the primary testing laboratory for impurities wt%, within 25% for impurities from 0.1-1.0 wt%. Documentation in support of precision studies should include the standard deviation, relative standard deviation, coefficient of variation, and the confidence interval. Reproducibility is not normally expected if intermediate precision is performed.

4.3 Specificity

For chromatographic methods, developing a separation involves demonstrating specificity, which is the ability of the method to accurately measure the analyte response in the presence of all potential sample components. The response of the analyte in test mixtures containing the analyte and all potential sample components (placebo formulation, synthesis intermediates, excipients, degradation products, process impurities, etc.) is compared with the response of a solution containing only the analyte.

Other potential sample components are generated by exposing the analyte to stress conditions sufficient to degrade it to 80-90% purity. For bulk pharmaceuticals, stress conditions such as heat (60-105°C, depends on the sample), light (600 FC of UV), acid (0.1N HCl), base (0.1N NaOH), and oxidant (3% H_2O_2) are typical (Table 3). For formulated products, heat, light, and humidity (70 to 80% RH) are often used. The resulting mixtures are then analyzed, and the analyte peak is evaluated for peak purity and resolution from the nearest eluting peak. If an alternate chromatographic column is to be allowed in the final method procedure, it should be identified during these studies. Once acceptable resolution is obtained for the analyte and potential sample components, the chromatographic parameters, such as column type, mobile-phase composition, flow rate, and detection mode, are considered set. An example of specificity criteria for an assay method is that the analyte peak will have baseline chromatographic resolution of at least 1.5 from all other sample components. If this cannot be achieved, the unresolved components at their maximum expected levels will not affect the final assay result by more than 0.5%. Examples of specificity criteria for an impurity method is that all impurity peaks that are 0.1% by area will have baseline chromatographic resolution from the main component peak(s) and, where practical, will have resolution from all other impurities.

TABLE 3 Suggested conditions for performing forced degradation studies

Stress conditions	Suggested sample treatment
Acidic	24h in 0.1N HCl solution
Basic	24h in 0.1N NaOH solution
Oxidative	24h in 3% H_2O_2 solution
Heat	1h in 60-105°C
Light	1h expose to UV lamp (200 – 300 nm)

4.4 Limit of Detection

The limit of detection (LOD) is defined, as the lowest concentration of an analyte in a sample that can be detected, not quantitated. It is a limit test that specifies whether or not an analyte is above or below a certain value. It is expressed as a concentration at a specified signal-to-noise ratio, usually three-to-one. The ICH has recognized the signal-to-noise ratio convention, but also lists two other options to determine LOD: visual non-instrumental methods and a means of calculating the LOD. Visual non-instrumental methods may include LOD determined by techniques such as thin layer chromatography (TLC). LOD may also be calculated based on the standard deviation of the response (SD) and the slope of the calibration curve(s) at levels approximating the LOD according to the formula: LOD =

3.3(SD/S). The standard deviation of the response can be determined based on the standard deviation of the blank, on the residual standard deviation of the regression line, or the standard deviation of y-intercepts of regression lines. The method used to determine LOD should be documented and supported, and an appropriate number of samples should be analyzed at the limit to validate the level.

4.5 Limit of Quantitation

The limit of quantitation (LOQ) is defined as the lowest concentration of an analyte in a sample that can be determined with acceptable precision and accuracy under the stated operational conditions of the method. Sometimes a signal-to-noise ratio of ten-to-one is used to determine LOQ. This signal-to-noise ratio is a good rule of thumb, but it should be remembered that the determination of LOQ is a compromise between the concentration and the required precision and accuracy. That is, as the LOQ concentration level decreases, the precision increases.

The ICH has recognized the ten-to-one signal-to-noise ratio as typical, and also, like LOD, lists the same two additional options that can be used to determine LOQ, visual non-instrumental methods and a means of calculating the LOQ. The calculation method is again based on the standard deviation of the response (SD) and the slope of the calibration curve(s) according to the formula: LOQ = 10(SD/S). Again, the standard deviation of the response can be determined based on the standard deviation of the blank, on the residual standard deviation of the regression line, or the standard deviation of y-intercepts of regression lines.

The method used to determine LOQ should be documented and supported, and an appropriate number of samples should be analyzed at the limit to validate the level. One additional detail should also be considered; both the LOQ and the LOD can be affected by the chromatography. Figure 2 shows how efficiency and peak shape can affect the signal-to-noise ratio. Sharper peaks result in a higher signal-to-noise ratio, resulting in lower LOQ and LOD.

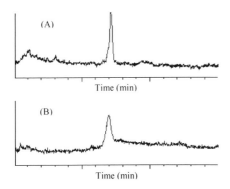

FIGURE 2 Effect of peak shape on detection limit and quantitation limit. (A) Waters Symmetry C18 column, quantitation limit, signal-to-noise = 11, (B) detection limit, signal-to-noise = 6.5.

4.6 Linearity and Range

Linearity is the ability of the method to elicit test results that are directly proportional to analyte concentration within a given range.

Range is the interval between the upper and lower levels of analyte (inclusive) that have been demonstrated to be determined with precision, accuracy and linearity using the method as written. The range is normally expressed in the same units as the test results obtained by the method. The ICH guidelines specify a minimum of five concentration levels, along with certain minimum specified ranges.

For assay, the minimum specified range is from 80-120% of the target concentration. For an impurity test, the minimum range is from the reporting level of each impurity to 120% of the specification. (For toxic or more potent impurities, the range should be commensurate with the controlled level.) For content uniformity testing, the minimum range is from 70-130% of the test or target concentration, and for dissolution testing, ±20% over the specified range of the test. In the literature it is often seen that a range, 25% – 150% or 200% of the nominal concentration of analyte is examined [14].

In practice the linearity study should be designed to be appropriate for the intended analytical method (Table 4). Acceptability of linearity data is often judged by examining the correlation coefficient and y-intercept of the linear regression line for the response versus concentration plot (see example of plot in Figure 4). A correlation coefficient of > 0.999 is generally considered as evidence of acceptable fit of the data to the regression line. The y-intercept should be less than a few percent of the response obtained for the analyte at the target level.

Linearity of progesterone (Figure 3) drug was studied in the concentration range 0.025-0.15 mg/ml (25-150% of the theoretical concentration in the test preparation, n = 3) and the following regression equation was found by plotting the peak area (y) versus the progesterone drug concentration (x) expressed in mg/mL: y = 3007.2x + 4250.1 (r^2 = 1.000) (Figure 4). The demonstration coefficient (r^2) obtained for the regression line demonstrates the excellent relationship between peak area and concentration of progesterone. The analyte response is linear over the range of 80 to 120% of the target concentration for progesterone.

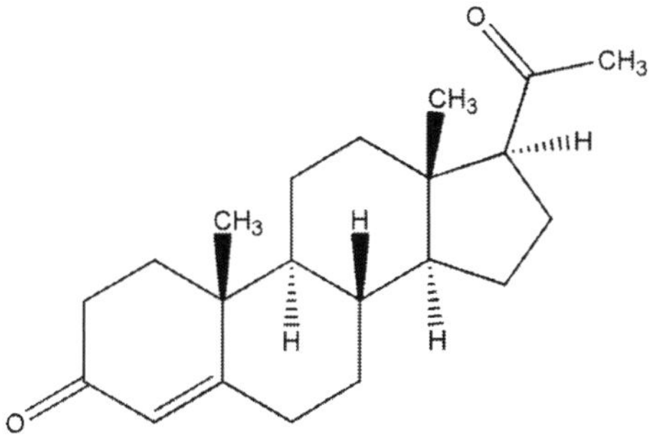

FIGURE 3 Chemical structure of progesterone.

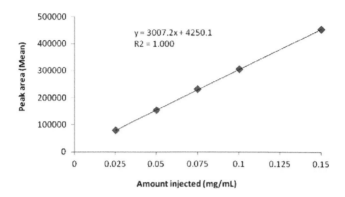

FIGURE 4 Amount injected versus peak area of progesterone standard to demonstrate linearity.

In addition, goodness of fit of data to the regression line may be evaluated by a procedure based on the residual sum of squares. Taking the regression line as the mean, a percent RSD is calculated for the data; normally this value should not be greater than 2.0%, but when evaluating this determination, the results of precision determinations should also be taken into account.

TABLE 4 Recommended validation ranges for linearity studies

Analysis categories	Typical range (%)	Recommended validation range (%)
Assay specifications for release	95 – 105	80 – 120
Assay specification for check	90 – 110	80 – 120
Content uniformity test	75 – 125	70 – 130
Assay of a preservative in a stability study	50 – 110	40 – 120
Determination of a degradant in a stability study	0 – 10	0 – 20

4.7 Robustness Studies

The robustness of a method is its ability to remain unaffected by small deliberate variations in method parameters. The robustness of a method is evaluated by varying method parameters such as percent organic solvent, pH of buffer in mobile phase, ionic strength, different HPLC columns (lots and / or suppliers), column temperature, flow rate etc. These method parameters may be evaluated one factor at a time or simultaneously as part of a factorial experiment [15]. As documented in the ICH guidelines, robustness should be considered early in the development of a method. In addition, if the results of a method or other measurements are susceptible to variations in method parameters, these parameters should be adequately controlled and a precautionary statement included in the method documentation. An example of robustness criteria is that the effects of the following changes in chromatographic conditions should be determined (Table 5): percent organic solvent in mobile phase adjusted by (±2 to 5%), mobile-phase pH adjusted by (up to ±0.5 pH units), gradient slope (by 2 to 5%, if appropriate), flow rate (±0.2 mL/min), wavelength (±5 nm), column temperature adjusted by (±1 to 5°C) and buffer ionic strength level of additives in the mobile phase. If these changes are within the limits that produce acceptable chromatography, they will be incorporated in the method procedure.

TABLE 5 Chromatographic parameter recommended for robustness study

Parameter	Test conditions
Percent organic solvent in mobile phase	Adjusted by ±2 to 5%
Mobile phase pH	Adjusted by (up to ±0.5 pH units
Gradient slope	by 2 to 5%, if appropriate
Flow rate	±0.2 mL/min
UV wavelength	±5 nm
Analytical column temperature	adjusted by ±1 to 5°C

5. ANALYTICAL PERFORMANCE PARAMETER REQUIRED FOR ASSAY VALIDATION

Both the USP and the ICH recognize that is it not always necessary to evaluate every analytical performance parameter. The type of method and its intended use indicates which parameters need to be investigated, as can be seen in Table 6. The USP divides analytical methods into four separate categories: Quantitation of major components or active ingredients; Determination of impurities or degradation products; Determination of performance characteristics (e.g., dissolution, drug release); Identification tests.

For assays in category 1, LOD and LOQ evaluations are not necessary because the major component or active ingredient to be measured is normally present at high levels. However, since quantitative information is desired, all of the remaining analytical performance parameters are pertinent. Assays in category 2 are divided into two sub-categories: Quantitative and limit tests. If quantitative information is desired, a determination of LOD is not necessary, but the remaining parameters are required. The situation reverses itself for a limit test. Since quantitation is not required, it is sufficient to measure the LOD and demonstrate specificity and ruggedness.

The parameters that must be documented for methods in USP assay category 3 are dependent upon the nature of the test. Dissolution testing, for example, falls into this category. The ICH treats analytical methods in much the same manner, as shown in Table 7. USP categories 1 and 2 match the ICH categories of assay and impurity testing, respectively, and the corresponding discussion above still applies.

TABLE 6 USP characteristics required for assay validation

Analytical Performance Parameter	Assay Category 1	Assay Category 2 Quantitative	Limit Tests	Assay Category 3	Assay Category 4
Accuracy	Yes	Yes	*	*	No
Precision	Yes	Yes	No	Yes	No
Specificity	Yes	Yes	Yes	*	Yes
LOD	No	No	Yes	*	Yes
LOQ	No	Yes	No	*	No
Linearity	Yes	Yes	No	*	No
Range	Yes	Yes	*	*	No

*May be required, depending on the nature of the specific test.

The ICH has not yet chosen to specifically address methods for performance characteristics (USP category 3), however, but has instead addressed analytical methods for compound identification. In this ICH category, it is only necessary to prove that the method is specific for the compound being identified.

TABLE 7 ICH validation characteristics

Analytical Performance Characteristics	Identification	Impurity Testing		Assay
		Quantitative	Limit Tests	
Accuracy	No	Yes	No	Yes
Precision				
Repeatability	No	Yes	No	Yes
Intermediate precision	No	Yes	No	Yes
Specificity	Yes	Yes	Yes	Yes
LOD	No	No	Yes	No
LOQ	No	Yes	No	No
Linearity	No	Yes	No	Yes
Range	No	Yes	No	Yes

6. CONCLUSION

Validation is a constant, evolving process starting before an instrument is placed on-line, and continues long after method development and transfer. A well-defined and documented validation process provides regulatory agencies with evidence that the system and method is suitable for its intended use. By approaching method development, optimization and validation in a logical, stepwise fashion, laboratory resources can be used in a more efficient and productive manner. I hope that I have provided a complete guide to help you to understand of how to perform an analytical method validation that generates both quality and meaningful data that meets all FDA, USP and ICH validation requirements.

REFERENCES

1. Reviewer Guidance, Validation of Chromatographic Methods, Center for Drug Evaluation and Research, Food and Drug Administration, 1994.
2. Guideline for Submitting Samples and Analytical Data for Methods Validation. Food and Drug Administration, February 1987.
3. U.S. Pharmacopeia 34, Validation of Compendial Methods, Section (<1225>), United States Pharmacopeal Convention, Inc., Rockville Maryland, 2011.
4. International Conference on Harmonisation (ICH), Q2(R1): Validation of Analytical Procedures: Text and Methodology, Geneva, Nov. 2005.
5. U.S. FDA HSS, 21 CFR Part 211, current Good Manufacturing Practice of certain requirements for finished pharmaceuticals, Proposed Rule, May 1996.
6. U.S. Pharmacopeia 34, Chromatography, Section (<621>), United States Pharmacopeal Convention, Rockville Maryland, 2011
7. Virlichie, J. L. and Ayache, A., Ruggedness Test and it's Application for HPLC Validation. S. T. P. *Pharma Pratiques*, 5(1) (1995) 49.
8. Green, M. J., A Practical Guide to Analytical Method Validation. *Anal. Chem.* 68 (1996) 305A.
9. R. Cassidy, M. Janoski, *LC/GC*, 10(9) (1992) 692.
10. Maxwell, W. and Sweeney, J., Applying the Validation Timeline to HPLC System Validation. *LC/GC* 12(9), (1994) 678.
11. M. P. Richards, J. H. Beattie, *J. Chromatogr.* 648 (1993) 549.
12. International Standard ISO/IEC 17025: General requirements for the Competence of Testing and Calibration Laboratories, 2000.

13. Draft Guidance for Industry: 21 CFR Part 11; Electronic Records; Electronic Signatures, Validation, U.S. FDA (August 2001).

14. Guidelines for Collaborative Study Procedure to Validate Characteristics of a Method of Analysis, *J. Assoc. Off. Anal. Chem.* 72, 694-704 (1989).

15. Virlichie J., L. Ayache A. S.T.P. *Pharma Pratiques* 5 (1995) 37.

4 SYSTEMATIC STRATEGIES IN HPLC METHOD DEVELOPMENT AND VALIDATION IN PHARMACEUTICAL ANALYSIS

ABSTRACT

This paper concerns the systematic strategies in high-performance liquid chromatography (HPLC) method development and validation. Currently, there is no single source and/or a lack of detailed recommendations on method validation that helps analytical scientists to perform validation in a systematic manner in pharmaceutical industry. Therefore industry depends on the scientist's knowledge and experience to develop simple and efficient methods of analysis. This is why much effort in this paper is focused on the development of validated methods in the pharmaceutical industry. Therefore, guidelines were given for analytical development and validation in this field; the methyl p-hydroxybenzoate and propyl p-hydroxybenzoate was used as model. In addition, a rapid method using HPLC coupled with photodiode array detection was developed and validated for the simultaneous determination of methyl p-hydroxybenzoate and propyl p-hydroxybenzoate in liquid pharmaceutical sample.

Keywords: HPLC, Method development, Method validation, FDA, USP, ICH

Methyl *p*-hydroxybenzoate, Propyl *p*-hydroxybenzoate, Liquid pharmaceutical formulation

1. INTRODUCTION

Method validation issues are especially important in the analytical field when quantification is made. When a test method has been developed it is important to validate it to confirm that it is suitable for its intended use. The method validation is today an essential concern in the activity of analytical chemistry laboratories. Validation has received considerable attention in the literature. The U.S. FDA [1] describes in Section 211.165 (e) under cGMP, industries should include validation parameters such as accuracy, sensitivity, specificity, and reproducibility in their test method validation and documented. Validation and documentation may be accomplished in accordance with Section 211.194(a). These requirements include a statement of each method used in testing the sample to meet proper standards of accuracy and reliability, as applied to the tested product. The U.S. FDA has also proposed industry guidance for Analytical Procedures and Methods Validation [2]. International Organization for Standardization/International Electrotechnical Commission (ISO/IEC) 17025 includes a chapter on the validation of methods [3] with a list of nine validation parameters. The ICH has introduced a guideline [4] on the validation of analytical procedures. The document includes definitions for eight validation characteristics. The U.S. Pharmacopoeia (USP) has published specific guidelines for compendial method validation [5] and more recently has introduced a new chapter on verification of compendial procedures [6].

Method validation is a critical step for any product release for marketing authorisation. The literature contains diverse approaches to performing method validation [7-22]. Many analytical methods appearing in the literature have not been through a thorough validation exercise and thus should be treated with caution until full validation has been carried out. Also there is no method validation reported with cGMP considerations. Validation of a new method is a costly and very time-consuming exercise. However the result of not carrying out method validation could result in litigation, failure to get product approval, costly repeat analysis and loss of business and market share [23].

Currently, there is no completely worldwide single source or final guideline on method validation that helps analysts to perform validation in a systematic manner and most importantly under cGMP considerations. Therefore industry depends on the analyst's knowledge and experience to develop simple and efficient methods of analysis. The ICH achieved a great deal in harmonising the definitions of the required validation characteristics and their basic requirements. However, they provide only a basis for a general discussion of the validation parameters, their calculation and interpretation. Also ICH has not explained step-by-step approaches and most importantly cGMP considerations that required during method validation to meet regulatory requirements. This impacts regulatory submissions.

The first section of this paper deals with systematic development and validation of analytical methods for pharmaceutical analysis. The second section reviews and demonstrates practical approaches to method validation with reference to simultaneous determination of methyl *p*-hydroxybenzoate and propyl *p*-hydroxybenzoate in a commercially available liquid pharmaceutical sample using HPLC.

The wide variety of equipment, columns, eluent and operational parameters makes the process of HPLC method development seems complex. HPLC method development is drawn by the nature of the analytes and generally follows the following steps: *step one*, selection of the HPLC method and initial system; *step two*, selection of initial conditions; *step three*, selectivity optimisation; *step four*, system optimisation; *step five*, method validation. Depending on the overall requirements and nature of the sample, analytes, etc. not all of these steps will be necessary for all HPLC analysis. For example, a satisfactory separation may be found during step 2, in which case step three and step four may not be required. The extent to which method validation (step five) is investigated will also depend on the use of the end analysis e.g. a method required for quality control will require more validation than one developed for a one off analysis. One is overriding consideration when developing a HPLC method Keep it simple. Try the most common columns and stationary phases first. Thoroughly investigate binary mobile phases before going on to ternary, etc. Think of the factors which are likely to be significant in achieving the desired resolution (e.g. the mobile phase composition in the most powerful way of optimising selectivity whereas temperature has only a minor effect; therefore don't waste time trying to achieve small selectivity changes by attending the separation temperature when you could be altering the mobile phase composition. pH will only significantly affect the retention of weak acids and bases, etc.). The objective of this article is to present a simple systematic approach to HPLC methods development beginning with sample preparation and finishing with practical analytical methods validation.

2. HPLC METHOD DEVELOPMENT

2.1 Step 1 - Selection of the HPLC Method and Initial System

When developing an HPLC method, the first step is always to consult the chromatographic literature to find out if anyone else has done the separation, and how they did it. This will at least give you an idea of the conditions that are needed, and may save you having to do a great deal of experimental work. At the outset, a HPLC system needs to be selected which has a high probability of being able to meet the requirements of the end analysis. For example, if the sample includes polar analytes then reverse phase HPLC would be a good choice as there would be high probability that conditions could be found which would afford both adequate retention and resolution. Normal phase HPLC would be much less likely to be feasible. Factor selection of the following is required.

Sample preparation: Does the sample require dissolution, filtration, extraction, pre-concentration, clean up. Is chemical derivatization required to assist detection sensitivity or selectivity.

Types of chromatography: Reverse phase is the choice for the vast majority of samples, but if acidic or basic analytes are present then reverse phase ion suppression (for weak acids or bases) or reverse phase ion pairing (strong acids or bases) should be used. The stationary phase should be a C_{18} bonded phase (unless there are good reasons for using a different material). For low to medium polarity analytes normal phase HPLC is a potential candidate particularly if separation of isomers (e.g. enantiomers) is required. Cyano bonded phases are easier to work with than plain silica for normal phase separations. For the analysis of inorganic anions and cations ion exchange chromatography is usual. For the analysis of high molecular weight compounds (e.g. >ca. 2000) then size exclusion chromatography would normally be considered.

Gradient HPLC: Gradient HPLC is only a requirement when you have either: A) complex samples with a large number of components (ca. >20 - 30). Gradient HPLC is more suitable for complex samples as the maximum number of peaks that can be resolved with a given resolution is much higher than in isocratic HPLC. This is a result of the roughly constant peak width that is observed in gradient HPLC (in isocratic HPLC peak width increases in proportion to retention time). B) Samples containing analytes with a wide range of retentivities which would, under isocratic conditions, afford chromatograms with capacity factors outside of the normally-acceptable range of 0.5 - 10 - 15. Use of gradient HPLC will also result in greater sensitivity particularly for analytes with longer retention times. This, again, is the result of the more constant peak width encountered in gradient HPLC (for a given peak area, peak height is inversely proportional to peak width). Reverse phase gradient HPLC is commonly used in the analysis of peptides and small proteins using an acetonitrile - water mobile phase containing 1% trifluoroethanoic acid. Gradient HPLC is an excellent method for initial exploration of samples.

Column dimensions: For most samples (unless they are very complex), the use of short columns (e.g. 15 or 10 cm) can be recommended to reduce method development time. Shorter columns afford shorter retention and equilibration times all other factors being equal. A flow rate of 1 - 1.5 ml/min should be used initially. Packing particle size should be 5 or 3 μm.

Detectors: Do the analytes have chromophores to enable UV detection. Do you require more selective/sensitive detection (Table 1). What detection limits is necessary. Will the sample require chemical derivatisation to enhance detectability and or improve the chromatography. For trace analysis fluorescence or electrochemical detector are preferred. For preparative HPLC refractive index is preferred because it can handle high concentrations without overloading the detector.

TABLE 1 HPLC detectors and their attributes comparison

Detector	Analyte/Attributes	Sensitivity
UV-Vis absorbance	Specific: compounds with UV chromophores	ng
Diode array	Specific: same as UV-Vis detectors but also provides UV spectra	ng
Fluorescence	Specific: compounds with native fluorescence or with fluorescence tag	fg-pg
Refractive index	Universal: polymers, sugars, triglycerides, organic acids, excipients	0.1-$10\mu g$
Evaporative light-scattering	Universals: non-volatile or semi-volatile compound, compatible to gradients	Low ng
Electrochemical	Specific: electro-active compounds	pg
Conductivity	Specific: anions and cations, organic acids, surfactants	ppm-ppb
Radioactivity	Specific: radio-active labelled compounds	Low levels
Mass spectrometry	Both universal and specific: definitive, identification	fg-pg- ng

Selecting UV wavelength: Optimum wavelength requires knowledge of the UV spectra, for the greatest sensitivity use the λ_{max}. However, normally it is a compromise wavelength for detecting all the sample components, when possible, UV wavelength below 200nm should be avoided because detector noise increases in the region. Higher wavelengths give greater selectivity.

Selecting fluorescence wavelength: Excitation wavelength scanned first, this locates the excitation maximum i.e. wavelength which gives the maximum emission intensity. The excitation is set to the maximum value then the emission is scanned to locate the emission intensity. Selection of the initial system could therefore be based on assessment of the nature of sample and analytes together with literature data, experience, expert system software and empirical approaches.

2.2 Step 2 - Selection of Initial Conditions

The aim in this step is to find conditions such that all analytes are adequately retained i.e. no analyte has a capacity factor of less than 0.5 (pour retention could result in the analyte peak overlapping with solvent and other unretained peaks) and no analyte has a capacity factor greater than 10 -15 (excessive retention leads to long analysis time and broad peaks with poor detectability). Factor selection of the following is required.

Mobile phase solvent strength: The mobile phase solvent strength is a measure of its eluting powers i.e. its ability to pull analytes off the column. It is generally controlled by the concentration of the strong solvent (i.e. the solvent with the highest solvent strength e.g. in reverse phase HPLC with aqueous mobile phases the strong solvent would be the organic modifier; in normal phase HPLC, the strong solvent would be the most polar one such as dichloromethane in a hexane/dichloromethane system). So the aim is to find the correct concentration of the strong solvent. In fact, with many samples, there will be a range of solvent strength that can be used while still maintaining capacity factors within the above range. Other factors (e.g. mobile phase pH with acids and bases, presence of ion pairing reagents with strong acid/bases) may also affect the overall retention of analytes and should be considered.

In reversed-phase HPLC, selecting the proper buffer pH is necessary to reproducibly separate ionizable compounds. Buffers are needed when the analyte is ionizable under reversed-phase

HPLC conditions or the sample solution is outside the pH 2 and 7.5 range. Analytes ionizable under reversed-phase HPLC conditions often have amine or acid functional groups with pK_a between 1 and 11. A correctly chosen buffer pH will ensure that the ionizable functional group is in a single form, whether ionic or neutral. Table 2 sumarises the common buffers for HPLC and their respective pK_a and UV cutoffs.

TABLE 2 Common HPLC buffers and their respective pK_a and UV cutoff

Buffer	pK_a	UV cutoff (nm)
Trifluoroacetic acid[1]	0.3	210
Phosphate	2.1, 7.2, 12.3	190
Citrate	3.1, 4.7, 5.4	225
Formate[1]	3.8	200
Acetate	4.8	205
Carbonate[1]	6.4, 10.3	200
Tris(hydroxymethyl) aminomethane	8.3	210
Ammonia[1]	9.2	200
Borate	9.2	190
Diethylamine	10.5	235

[1]volatile buffer systems that are MS compatible

It is necessary to know the pK_a value of the analyte before buffer pH can be selected. A buffer pH 2 units above or below pK_a is recommended for good peak shape and reproducibility. From the Henderson-Hasselback equation (1), it can be determined that 99% of an analyte is in a single form if the solution pH is 2 units above or below the analye's pK_a value.

$$pH = pKa + \log [A^-] / [HA] \quad (Eq. 1)$$

Gradient HPLC: With samples containing a large number of analytes (ca. >20 - 30) or with a wide range of analyte retentivities gradient elution will be necessary to avoid excessive retention of analytes.

Determination of initial conditions: The recommended method involves performing two gradient runs differing in only the gradient run time. You should be using a binary system based on either acetonitrile/water (or aqueous buffer) or methanol/water (or aqueous buffer).

2.3 Step 3 - Optimisation of Selectivity

The aim of this step is to achieve adequate selectivity (peak spacing). Factor; mobile phase composition and stationary phase composition. In order to minimise the number of trial chromatograms involved you need to be careful to examine only the parameters that are likely to have a significant effect on selectivity in the optimisation. In order to select these, the nature of the analytes must be considered. For this, it is useful to categories analytes (Table 3) into a few basic types.

TABLE 3 Basic types of analytes in HPLC

Analyte type	Characteristics
Neutral	No significantly acidic or basic functional groups
Weak acid	Has carboxylic acid function or phenolic -OH
Weak base	Aromatic amine
Strong base	Non-aromatic amine

Once the analyte types are identified the relevant optimisation parameters may be selected (see Table 4). Note that optimisation of mobile phase parameters is always considered first as this is much easier and convenient than stationary phase optimisation.

TABLE 4 Optimisation parameters in HPLC

Analytes	HPLC method	Optimise
Neutral	Reversed-phase	Solvent strength, solvent type
Weak acids and/or weak bases	Ion suppression	pH, solvent strength, solvent type
Strong acid and/or strong bases	Ion pairing	Ion pairing reagent concentration, pH Solvent strength, solvent type
Inorganic anions/cations	Ion exchange	Eluting ion concentration

Optimisation of selectivity in gradient HPLC: Initially you should attempt to optimise gradient conditions using a binary system based on either acetonitrile/water (or aqueous buffer) or methanol/water (or aqueous buffer). If you are seeing a serious lack of selectivity, you should initially investigate the use of a different organic modifier.

2.4 Step 4 - Optimisation of System Parameters

System optimisation is used to find the desired balance between resolution and analysis time after satisfactory selectivity has been achieved. The parameters involved in system optimisation include: column dimensions, column-packing particle size and flow rate. These parameters may be changed without affecting capacity factors or selectivity.

2.5 Step 5 – Method Validation

Proper validation of analytical methods is of particular importance in relation to the analysis of medicines where assurance of the continuing efficacy and safety of each batch manufactured relies solely on the determination of quality. Ability to control this quality is clearly dependent upon the ability of the analytical methods, as applied under well-defined conditions and at established level of sensitivity, to give a reliable demonstration of all deviation from target criteria such as content of active principle and preservatives, polymorphic forms, permitted levels of impurities etc. were any deviation to occur. Validation of analytical methods is now required by the various regulatory authorities for marketing authorisations, and guidelines have been published. It is important to isolate analytical methods validation from the selection and development of the method. Method selection is the first step in establishing an analytical method. During method selection, the analyst must consider what it is to be measured and with what accuracy and precision. Method development and validation may go on simultaneously, but they are two different processes both "downstream" of method selection. Analytical methods used in quality control should ensure an acceptable degree of confidence that results of the analyses of raw materials, excipients, intermediates, bulk products or finished and viable. Before a test procedure is validated it must decided what criteria are to be used in that validation. Validation of analytical methods involves examining a number of different criteria. Analytical methods should be used within a good manufacturing practice (GMP) and good laboratory practice (GLP) regulated environment and must be developed using the protocols set out in the ICH guidelines (Q2R1). The FDA methods validation guidelines and USP both refer to ICH guidelines. Method validation must have a written and approved protocol and trained analysts prior to their initiation.

In this paper it is intended to review and demonstrate practical approaches to analytical method validation with reference to an HPLC assay of methyl p-hydroxybenzoate (MHB) and propyl p-hydroxybenzoate (PHB) in liquid pharmaceutical formulation. MHB and PHB (0.05 to 0.25%), alone or in combination with other esters of p-hydroxybenzoic acid or with other antimicrobial agents are used as a preservatives in cosmetic, food and pharmaceutical formulations [24].

3. STRATEGIES FOR THE METHOD DEVELOPMENT AND VALIDATION

Everyday many analytical scientists face the need and challenge to develop and validate new analytical methods. Whereas individual's approaches exhibit considerable diversity, following the systematic approach given in Figure 1 can simplify the process. In the feasibility phase, the analyst will determine whether the assigned task can be successfully accomplished by using available resources. Research is defined as the activity aimed at discovering new knowledge on the compound in hopes that such information will be useful in developing a new method. Development phase is the translation of research findings into a new analytical method and the systematic use of knowledge or understanding gained from research, directed towards the analytical method, including the design and development of prototypes and processes. The development phase must include studies of robustness, system suitability and stability of analytical solutions as well as mobile phase. In optimization study, developed method can be further improved to gain greater confidence on the generation of analytical data. The current developed approach emphasis the allocation of greater resources during the development and optimization phases. This allows the analyst to have more confidence on the quality of data generated and therefore considerably reduces the resources that are required for the process of validation. The purpose of the characterization study is to determine reliable method performance limits from the analytical performance characteristics and set acceptance criteria for the method validation. As a best practice, the characterization protocol needs to be written and approved before execution. Prior to execution of the protocol, it is necessary that the analytical system itself is adequately designed, maintained, calibrated and validated. In all cases proper documentation should be archived to support the validation process. All personnel involved in the characterization protocol activities must be trained prior to performing their function. On completion of the characterization study, the results/data should be critically assessed from a statistical point of view.

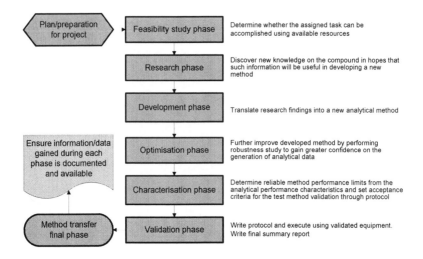

FIGURE 1 Systematic approach in method development and validation process.

Validation is the last and critical step for the success of the whole method development project. If the validation fails, it can be seen as a wasted resource and inevitably can delay the product release date. Here, validation protocol needs to be written and approved by an appropriate cross-functional team. Upon successful completing the validation the data should be statistically analyzed and its acceptance criteria by appropriate experts in order to test its validity. Method transfer plays an important role in expediting drug candidates through development stages. Method transfer is not a trivial task and requires careful planning and constant communication between the laboratory personnel involved in the transfer process. Method transfer could occur within the same organization or between pharmaceutical companies and analytical service providers. To have a successful transfer, the analytical method itself must be robust and the equipment differences between the delivering and receiving parties should be carefully evaluated.

4. PRACTICAL HPLC METHOD VALIDATION

This section reviews and demonstrates practical approaches to method validation with reference to simultaneous determination of methyl p-hydroxybenzoate and propyl p-hydroxybenzoate (Figure 2) in a commercially available liquid pharmaceutical sample using HPLC.

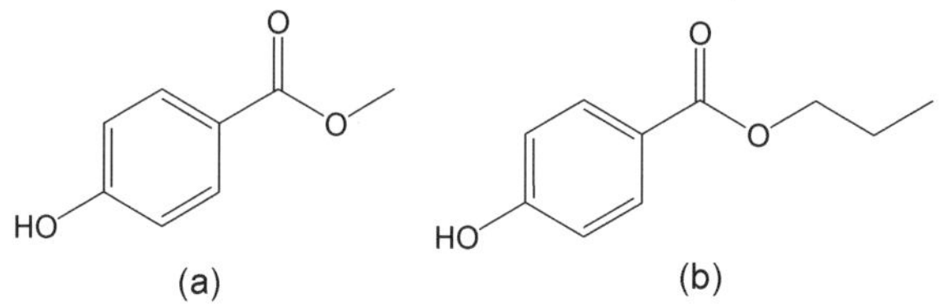

FIGURE 2 Structures of: (a) methyl p-hydroxybenzoate and (b) propyl p-hydroxybenzoate.

4.1 Materials and Methods

Methyl p-hydroxybenzoate (purity > 99%) and propyl p-hydroxybenzoate (purity > 99%) used were purchased from Sigma (St. Louis, MO). Potassium dihydrogen phosphate, sodium hydroxide (analytical reagent grade) and methanol (HPLC grade) were obtained from Merck (Darmstadt, Germany). Distilled water was de-ionised by using a Milli-Q system (Millipore, Bedford, MA).

The Knauer HPLC system (Berlin, Germany) equipped with a model 1000 LC pump, model 3950 autosampler, model 2600 photodiode-array (PDA) detector and a vacuum degasser was used. The data were acquired via Knauer ClarityChrom Workstation data acquisition software. The mobile phase was an aqueous solution of 50% (v/v) methanol containing 0.2 M potassium dihydrogen phosphate, adjusted to pH 6.05 $\pm$0.05 with 1 M NaOH. The flow rate was set to 1.0 mL/min. The injection volume was 20 μL and the detection wavelength was set at 258 nm. HPLC analysis was performed isocratically at ambient temperature using a Lichrosorb C_8 (150 $\times$ 4.6 mm, 5 μm particle size) column (Jones Chromatography, Hengoed, UK).

4.2 Standard and Sample Preparation

An accurately weighed amount (40 mg) of MHB and 20 mg of PHB standard were placed in a 100 mL volumetric flask and dissolved in methanol (stock solution 1). A 10 mL aliquot of stock solution 1 was transferred to a second 100 mL volumetric flask and diluted to volume with mobile phase yielding a final concentration of 0.04 mg/mL and 0.02 mg/mL, respectively.

Approximately 2.5 g of liquid pharmaceutical sample was accurately weighed, added to a 50 mL volumetric flask and then diluted in mobile phase.

4.3 Method Development

The chromatographic separation of MHB and PHB was carried out in the isocratic mode using a mixture of 50% methanol in potassium phosphate buffer pH 6.05 ±0.05 (50:50, v/v) as mobile phase. The column was equilibrated with the mobile phase flowing at 1.0 mL/min for about 20 min prior to injection. The column temperature was ambient. The PDA UV detector was set at 200-400 nm and 258 nm was chosen as the optimal wavelength for maximum detection sensitivity for both preservatives. Additionally, preliminary system suitability, precision, linearity, stability of solutions and robustness studies performed during the development of the method showed that the 20 µL injection volume was reproducible and the peak response was significant at the analytical concentration chosen. Chromatograms of the resulting solutions (standard and sample) gave very good peak shapes and resolution (Figure 3 and 4). In Table 1, the retention times are reported.

Robustness of the method was evaluated by the analysis of MHB and PHB under different experimental conditions such as changes in the composition of the mobile phase, flow rate, buffer pH and columns from different batches. The percentage of methanol in the mobile phase was varied ±2%, buffer pH (±0.5 pH units) and the flow rate was varied ±0.2 mL/min. Their effects on the retention time asymmetry factor, recovery and repeatability were studied. Deliberate variation of the method conditions had no significant effect on assay data or on chromatographic performance, indicating the robustness of the method and its suitability for routine use and transfer to other laboratories.

System suitability testing is essential for the assurance of the quality performance of the chromatographic system. The amount of testing required will depend on the purpose of the test method. For dissolution or release profile test methods using an external standard method, capacity factor, tailing factor and RSD are minimum recommended system suitability tests. For acceptance, release, stability, or impurities / degradation methods using external or internal standards, capacity factor, tailing factor, resolution and RSD are recommended as minimum system suitability testing parameters.

In this study, system suitability testing was performed to determine the accuracy and precision of the system from six replicate injections of solutions containing 0.04 and 0.02 mg/mL of MHB and PHB, respectively. The percent relative standard deviations (RSD) of the peak area were found to be less than 0.14% for both preservatives. The retention factor (also called capacity factor, k) was calculated using the equation $k = (t_r \, / \, t_0) -1$, where t_r is the retention time of the analyte and t_0 is the retention time of an unretained compound; in this study, t_0 was calculated from the first disturbance of the baseline after injection and capacity factor value was obtained 2.82 for MHB peak and 6.32 for PHB peak.

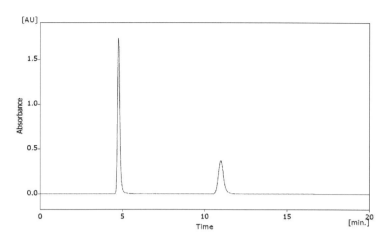

FIGURE 3 HPLC chromatogram of reference standard: MHB, RT, 4.98 (min) and PHB, RT, 11.85 (min).

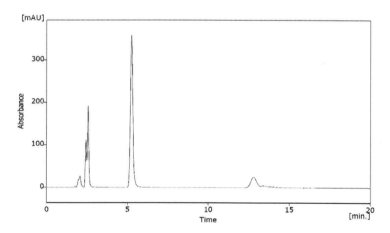

FIGURE 4 HPLC chromatogram of the sample: MHB, RT, 5.01 (min) and PHB, RT, 11.88 (min).

The plate number (also known as column efficiency, N) was calculated as $N = 5.54 \, (t_r / w_{0.5})^2$ where $w_{0.5}$ is the peak width at half peak height. In this study, the theoretical plate number was 4093 and 4862 for MHB and PHB, respectively. Resolution is calculated from the equation $R_s = 2(t_2 - t_1) / (t_{w1} + t_{w2})$. Where t_1 and t_2 are retention times of the first and second eluted peas, respectively, and t_{w1} and t_{w2} are the peak widths. The resolution for PHB peak was 13.89. The asymmetry factor was calculated using the USP method. The peak asymmetry value for each MHB and PHB peaks were 1.12 and 1.25, respectively. The proposed method met these requirements within the USP and U.S. FDA accepted limits (Table 5) [25, 26].

TABLE 5 System suitability parameters and results obtained for MHB and PHB preservatives

Parameter	Acceptance criteria	Results	
		MHB	PHB
Retention time (min)	-	4.98	11.85
Injection repeatability[1]	$N = 5$ for RSD $\leq 2\%$	0.10	0.13
Peak asymmetry	≤ 2.0	1.12	1.25
Capacity factor	> 2.0	2.82	6.32
Resolution	> 2.0	-	13.89
Theoretical plates	> 2000	4093	4862

[1]Six replicate injections, peak area, RSD (%)

4.4 Stability of Analytical Solutions

Test solutions of three batches of MHB and PHB were prepared using the conditions cited in Section standard preparation. They were chromatographed at the beginning and after 24h. The stability of MHB, PHB and the mobile phase were calculated by comparing area response and area percent of two standards at 0.04 and 0.02 mg/mL over time. The solutions were stable during the investigated 24h and the RSD was <1.0% for peak area and height. Standard solutions stored in a capped volumetric flask on a laboratory bench under normal lighting conditions for 24h, were shown to be stable with no significant change in MHB and PHB concentration over this period (Table 6). This is indicated (0.4% changes in area between T = 0h and T = 24h).

TABLE 6 Stability of MHB and PHB in solution (n = 5)

Time (h)	Area (RSD, %)	Height (RSD, %)	Recovery (%)	Percent of initial
MHB				
0	0.44	0.08	99.98	
24	0.38	0.10	99.88	99.89
PHB				
0	0.14	0.09	99.97	
24	0.21	0.11	99.87	99.90

4.5 Method Validation

Before validation activities begun, required validation parameters and associated acceptance criteria must be established, documented and approved. The level and extent of method validation requirements must take into account the stage of product development, source (new or established test methods), type (physical property, potency, impurity), and intended use (investigational, in-process/release, process or cleaning validation, stability) of the test method. Also before undertaking the task of method validation, it is necessary that the analytical system itself is adequately designed, maintained, calibrated and validated. The staff carrying out the validation experiments must be properly trained on equipment and validation protocol. In all cases proper validation documentation should be archived to support the validation process.

4.5.1 Linearity and range

Linearity was studied using seven solutions in the concentration range 0.010-0.070 mg MHB/mL and 0.005-0.035 mg PHB/mL, (n = 3). The regression equations were found by plotting the peak area (y) versus the MHB and PHB concentration (x) expressed in mg/mL. The correlation coefficient (r^2) obtained for the regression line demonstrates that there is a strong linear relationship between peak area and concentration of MHB and PHB (Table 7).

TABLE 7 Linearity assessment of the HPLC method for the assay of MHB and PHB

Methyl p-hydroxybenzoate			Propyl p-hydroxybenzoate		
Conc* (mg/mL)	Conc as % of 0.04 (mg/mL)	Area (mAU s) (n = 3)	Conc (mg/mL)	Conc as % of 0.02 mg/mL	Area (mAU s) (n = 3)
0.010	25	402	0.005	25	187
0.020	50	1146	0.010	50	646
0.030	75	1902	0.015	75	1089
0.040	100	2635	0.020	100	1511
0.050	125	3328	0.025	125	1987
0.060	150	4088	0.030	150	2428
0.070	175	4826	0.035	175	2907
y_{MP} = 73.507x - 322.14 (r^2 = 0.9999)			y_{PP} = 90.157x - 266.71 (r^2 = 0.9998)		

*Concentration

4.5.2 Accuracy/recovery studies

Accuracy of the method was evaluated by fortifying a MHB and PHB sample solutions with three known concentrations of reference standard 0.02, 0.04, 0.06 mg/mL and 0.01, 0.02, 0.03 mg/mL (50 - 150%), respectively. Percent recoveries were calculated form differences between the peak areas obtained for fortified and unfortified solutions. Mean recoveries (n = 3) for sample MHB and PHB were found to be 99.72% and 99.61%, respectively, as shown in Table 8.

TABLE 8 Recovery of MHB and PHB from samples with known concentrations

Sample #	Percent of nominal	Amount of analyte (mg/mL)		Recovery (%) and confidence limits
		Added	Found	
Methyl *p*-hydroxybenzoate				
1	50	0.025	0.0249	99.60 ± 0.10
2	100	0.045	0.0449	99.77 ± 0.13
3	150	0.055	0.0549	99.80 ± 0.11
Propyl *p*-hydroxybenzoate				
1	50	0.010	0.0100	100.0 ± 0.11
2	100	0.020	0.0199	99.50 ± 0.09
3	150	0.030	0.0298	99.33 ± 0.12

4.5.3 Precision studies

The precision of the method was investigated with respect to repeatability and intermediate precision. The repeatability (intra-day precision) of the method was evaluated by assaying six replicate injections of the MHB and PHB at 100% of test concentration 0.04 and 0.02 mg/mL, respectively on the same day. The (RSD, %) of the retention time (min) and peak area were calculated as shown in Table 9.

TABLE 9 Demonstration of the repeatability of the HPLC assay

Injection #	Methyl p-hydroxybenzoate		Propyl p-hydroxybenzoate	
	t_R (min)	Area (mAUs)	t_R (min)	Area (mAUs)
1	4.983	2226	11.850	941
2	4.983	2235	11.850	943
3	4.982	2236	11.851	947
4	4.983	2221	11.851	939
5	4.983	2226	11.850	942
6	4.983	2231	11.850	943
Mean (6)	4.983	2229	11.850	943
RSD (%)	0.008	0.262	0.004	0.283

Intermediate precision (inter-day precision) was demonstrated by two analysts using two instruments and evaluating the relative peak area percent data across the two instrument systems at three different concentration levels (50, 100 and 150%) that cover the assay method range of MHB (0.01-0.07 mg/mL) and (0.005-0.035 mg/mL) of PHB. The RSD across the systems and analysts were calculated from the individual relative peak areas mean values at the 50, 100 and 150% of the test concentration. The inter-day RSD ($n =$ 3) are given in Table 10. All the data are within the acceptance criteria of 2.0%.

4.5.4   Specificity/selectivity

HPLC-PDA/UV isoplot chromatogram (Figure 5) demonstrates a good separation of the MHB and PHB. The isoplot chromatogram data consist of UV absorption spectra from 200 to 400 nm for each point along the chromatogram. Also injections of the extracted placebo were performed to demonstrate the absence of interference with the elution of the MHB and PHB. These results demonstrate (Figure 6) that there was no interference from the other materials in the liquid formulation and, therefore, confirm the specificity of the method.

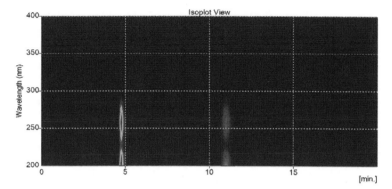

FIGURE 5 HPLC-PDA/UV isoplot chromatogram of MHB and PHB.

TABLE 10 Demonstration of the intermediate precision of the HPLC assay

Percent of nominal	HPLC Instrument 1			HPLC Instrument 2		
	50	100	150	50	100	150
Methyl *p*-hydroxybenzoate						
Analyst 1	1023[a]	2634	4987	1019	2636	4986
Analyst 2	1027[a]	2630	4982	1026	2629	4971
Mean instruments	1024	2633	4982			
Mean analysts	1025	2632	4984	1026	2632	4978
Instruments (%RSD)[b]	0.20	0.13	0.15			
Analysts (RSD, %)	0.28	0.11	0.07			
Propyl *p*-hydroxybenzoate						
Analyst 1	522	999	1502	530	992	1516
Analyst 2	518	996	1510	524	997	1511
Mean instruments	523	996	1509			
Mean analysts	520	998	1506	527	994	1513
Instruments (%RSD)	0.96	0.29	0.38			
Analysts (RSD, %)	0.54	0.21	0.37			

[a] Peak area (mAU s), $n = 3$
[b] Acceptance criteria (x) RSD, % < 2.0

Forced degradation studies were also applied to MHB and PHB reference standards at a concentration of 0.04 mg/mL and 0.02 mg/mL, respectively, to verify that none of the degradation products interfered with quantitation of the drug. Hydrolytic degradation was studied by heating the drug under reflux at 80°C in 0.1 M HCl and 0.1 M NaOH for 4h. The samples were then cooled to room temperature and neutralized. Oxidative degradation was studied by treating the drug with 3% H_2O_2 at room temperature (24 ±1°C) for 4h. Solutions containing 0.04 and 0.02 mg/mL of each degraded sample were prepared and injected in triplicate. A summary of the stress results for retention time, peak area and resolution are shown in Table 11. Under acidic and alkaline degradation hydrolysis conditions, the MHB content decreased and additional unknown polar peaks were observed near the solvent front. No degradation was observed under oxidative condition. This was further confirmed by peak purity analysis on a PDA UV detector. The MHB and PHB analytes obtained by acidic hydrolysis were well resolved from the additional peak, indicating the specificity of the method.

TABLE 11 Results of the stress conditions experiments

Stress conditions	Sample treatment	MHB		PHB		
		t_R (min)	Area (%)	t_R (min)	Area (%)	R_S
Reference	Fresh solution	4.983	70.2	11.850	29.9	13.8
Acid	1M HCl at 80°C for 4 h	4.733	66.7	10.817	30.7	7.4
Base	1M NaOH at 80°C for 4 h	4.750	68.4	10.867	30.8	7.6
Oxidative	3% H_2O_2 at room temperature for 24 h	4.982	69.9	11.841	30.01	13.7

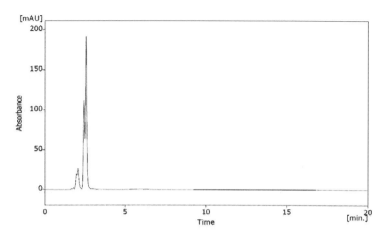

FIGURE 6 HPLC chromatogram of placebo without MHB and PHB in the sample formulation.

4.5.6 Limits of detection and quantitation

The limit of detection (LOD) is defined as the lowest concentration of an analyte in a sample that can be detected, not quantitated. It is expressed as a concentration at a specified signal-to-noise ratio, usually 3:1 [4]. The limit of quantitation (LOQ) is defined as the lowest concentration of an analyte in a sample that can be determined with acceptable precision and accuracy under the stated operational conditions of the method with a signal-to-noise ratio of 10:1. LOD and LOQ may also be calculated based on the standard deviation (σ) and the slope of the calibration curve(s) at levels approximating the LOD according to the equations:

$$LOD = 3.3\ \sigma/s$$
$$LOQ = 10\ \sigma/s$$

The standard deviation of the response can be determined based on the standard deviation of the blank, on the residual standard deviation of the regression line, or the standard deviation of y-intercepts of regression lines.

In this study, the LOD values for MHB and PHB were found to be 0.022 μg/mL (s/n = 3.05) and 0.009 μg/mL (s/n = 3.07), respectively. The LOQ values for MHB and PHB were found to be 10 μg/mL (s/n = 10.25) and s/n = 10.30 (5 μg/mL), respectively. The RSD for six injections of the LOQ solutions for MHB and PHB was < 2.0%.

5. SUMMARY AND CONCLUSION

In the present paper the detailed recommendation and rules given by the FDA, USP and ICH for analytical methods have been applied and modified to the field for analysis of pharmaceutical samples. This paper also contains guidelines for systematic method development and validation demonstrating proof of principle with a great potential to achieve excellent accuracy and precision of the quantification of results. HPLC coupled with PDA, for identification and/or quantification is today an established method in the pharmaceutical industry. In this paper, I also have developed and validated HPLC-PDA method that can be reliably used in simultaneously determination of liquid pharmaceutical samples containing methyl p-hydroxybenzoate and propyl p-hydroxybenzoate preservatives. I hope that I have provided a complete guide to help analysts to understand how to perform an analytical method validation under GMP environment that meets FDA, USP and ICH requirements.

REFERENCES

1. U.S. FDA. (2008) Code of Federal Regulations: 21 CFR 211, Current good manufacturing practice for finished pharmaceuticals.
2. U.S. FDA. (2000) Guidance for Industry, Analytical procedures and methods validation: Chemistry, Manufacturing, and Controls and Documentation.
3. ISO/IEC 17025. (2005) General requirements for the competence of testing and calibration laboratories.
4. ICH, Q2(R1). (2005) Validation of Analytical Procedures, Text and Methodology. Geneva, Switzerland.
5. U.S. Pharmacopoeia 34. (2011) *General Chapter* (1225), Validation of compendial methods, United States Pharmacopoeia Convention: Rockville, Maryland.
6. U.S. Pharmacopoeia 34. (2011) *General Chapter* (1226), Verification of compendial methods, United States Pharmacopeial Convention: Rockville, Maryland.
7 Crowther, J.B. (2001) Validation of pharmaceutical test methods. In: Handbook of Modern Pharmaceutical Analysis, Academic Press, California, 415.
8. Wilson, T.D. (1990) Liquid chromatographic methods validation for pharmaceutical products. *J. Pharm. Biomed. Anal.* 8: 389.
9. Clarke, G.S. (1994) The validation of analytical methods for drug substances and drug products in UK pharmaceutical laboratories. *J. Pharm. Biomed. Anal.* 12: 643.
10. Bressolle, F., Petit, M.B., Audran, M. (1996) Validation of LC and GC methods: application to pharmacokinetics. J. Chromatogr. B. 686: 3.
11. Carr, G.P., Wahlich, J.C. (1990) A practical approach to method validation in pharmaceutical analysis. *J. Pharm. Biomed. Anal.* 8: 613.

12 Green, M.J. (1996) A practical guide to analytical method validation. *Anal. Chem.* 68: 305A.

13. Trullols, E., Ruisanchez, I., Rius, F.X. (2004) Validation of qualitative analytical methods. *Trends Anal. Chem.* 23: 137.

14. Ermer, J. (2001) Validation in pharmaceutical analysis. Part 1: An integrated approach. *J. Pharm. Biomed. Anal.* 24: 755.

15. Daraghmeh, N., Al-Omari, M., Badwan, A.A., Jaber, A.M.Y. (2001) Determinaton of sildenafil citrate and related substances in the commercial products and tablet dosage form using HPLC. *J. Pharm. Biomed. Anal.* 25: 483.

16. Badea, I., Ciutaru, D., Lazar, Nicolescu, L. D., Tudose, A. (2004) Rapid HPLC method for the determination of paclitaxel in pharmaceutical forms without separation. *J. Pharm. Biomed. Anal.* 34: 501.

17. Mendez, A.S.L. Steppe, M. Schapoval, E.E.S. (2003) Validation of HPLC and UV spectrophotometric methods for the determination of meropenem in pharmaceutical dosage form. *J. Pharm. Biomed. Anal.* 33: 947.

18. Hokanson, G. C. (1994) A life cycle approach to the validation of analytical methods during pharmaceutical product development, Part I: The initial validation process, Pharm. Tech. 118.

19. Hokanson, G. C. (1994) A life cycle approach to the validation of analytical methods during pharmaceutical product development, Part II: Changes and the need for additional validation, *Pharm. Tech.* 92.

20. Vessman, J. (1996) Selectivity or specificity? Validation of analytical methods from the perspective of an analytical chemist in the pharmaceutical industry, *J. Pharm. Biomed. Anal.* 14: 867.

21. Huber, L., George, S. (1993) Diode-array detection in high-performance liquid chromatography, New York, Marcel Dekker, 4.

22. EURACHEM. (1998) The fitness for purpose of analytical methods: A laboratory guide to method validation and related topics.

23. Harvey, I.M., Baker, R. Woodget, B. (2002) Part 2, the analytical method. In: Series of tutorial lectures: Quality assurance in analytical sciences.

5 DEVELOPMENT AND VALIDATION OF A REVERSED-PHASE HPLC METHOD FOR 1,10-PHENANTHROLINE- 5,6-DIONE AND ANALYSIS OF ITS IMPURITIES BY HPLC-MS

ABSTRACT

A reversed-phase HPLC analytical method for the assay of 1,10-phenanthroline-5,6-dione (I) has been developed and validated. A C18 column (150 × 4.6 mm; 5 μm) was employed together with a mobile phase of methanol-water (50:50, v/v) containing 0.1% triethylamine. UV detection was performed at 254 nm. Dione (I) eluted as a spectrally pure peak resolved from its impurities allowing the method to be applied to the purity evaluation of samples obtained via two synthetic routes. In addition, 4,5-diazafluoren-9-one (V) was identified as the main impurity by employing the method in HPLC-MS mode with photodiode array UV detection.

Keywords: 1,10-Phenanthroline-5,6-dione; HPLC, HPLC-MS, Method development, Method validation, Synthetic impurities

1. INTRODUCTION

The phenanthroline-5,6-diones (I) – (III) are a useful class of heterocyclic o-quinone compounds (Figure 1). Historically, in the

1950s, they were first found to be of use because of their activity against protozoa, amoebae and bacteria [1]. The 4,7-isomer, also known as phanquone, proved to be the most active and was marketed as a treatment for amoebic dysentery under the tradename Entobex (Ciba Pharmaceuticals, now Novartis).

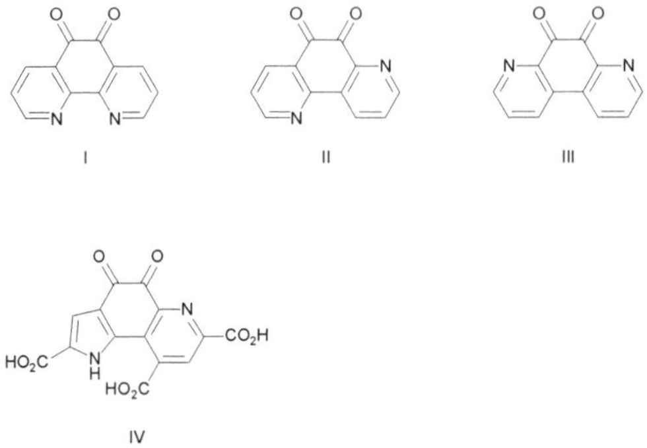

FIGURE 1 Heterocyclic o-quinone compounds.

Further interest [2,3,4] in compounds (I) – (III) was stimulated by the similarity of their structures and reactivity to that of the pyrroloquinoline quinone derivative methoxatin or PQQ (IV), which has been identified as a cofactor in various bacterial dehydrogenase enzymes. In particular, PQQ (IV) and the analogues (I) – (III) [4] together with transition metal complexes of (I) [5] have been suggested as catalysts for the regeneration of the cofactor NAD in the presence of oxygen. Latterly, electrochemical oxidation was substituted for aerial oxidation in these catalytic systems [6]. The quinones (I) – (III) are useful as NADH mediators in biosensors containing NAD-dependent enzymes for analytes such glucose or D-3-hydroxybutyrate [7,8]. Compound (I) has found widespread use as a starting material for the preparation of phenanthroline ligands for transition metals, particularly ruthenium.

The 1,10-isomer (I) was first reported by Smith and Cagle [9] in 1947 as a minor by-product (2% yield) of the nitration of 1,10-phenanthroline. Early attempts to prepare (I) – (III) via the direct oxidation of the parent phenanthroline using reagents such as chromic acid, selenium dioxide, vanadium pentoxide, iodic acid and periodic acid were unsuccessful [10]. However, Druey and Schmidt [11] did achieve the first syntheses of (I) – (III) in good yield by oxidizing the appropriate 5-methoxy phenanthroline derivative with a concentrated nitric/sulphuric acid mixture at 120°C. This route is also described in a patent [12] assigned to Ciba Ltd. A detailed method has been reported [13] for the case of the 4,7-isomer (III). The 1,10-isomer (I) can also be obtained by nitric acid oxidation of 5-amino-1,10-phenanthroline derived from the 5-nitro derivative [14,15]. Direct oxidation of 1,10-phenanthroline, as a cobalt complex, was first achieved in 1970 by adding potassium bromide to the nitric/sulphuric acid oxidizing medium [13]. Latterly, the use of the $KBr/HNO_3/H_2SO_4$ mixture was applied successfully in the preparation of (I) from 1,10-phenanthroline in one step by several groups [16,17,18]. Compounds (II) and (III) can be similarly prepared using more vigorous reaction conditions. The one-step oxidation and three-step cobalt complexation routes to (I) are illustrated in Scheme 1.

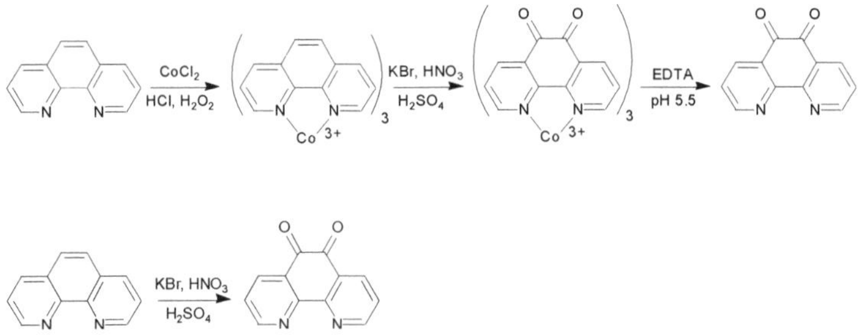

Scheme 1 Synthesis of 1,10-phenanthroline-5,6-dione (I).

We were interested in assessing the purity of the quinone (I) prepared by the two main synthetic routes [13,16] and identifying any impurities, where possible, to evaluate its suitability for use as a mediator to NADH in biosensors for diabetics. A number of potential impurities were expected from an examination of the literature concerning the reactivity of the phenanthroline-5,6-diones (Figure 2). Thus, compound (I) is known [11,19] to undergo alkaline decarboxylation to afford 4,5-diazafluoren-9-one (V). Here, (V) may be obtained during the neutralisation of highly acidic reaction mixtures in the preparation of (I). The use of $KBr/HNO_3/H_2SO_4$ reaction medium may result in the formation of 5-bromo-1,10-phenanthroline (VI) [16,20] and 5-nitro-1,10-phenanthroline (VII) [9,21] as by-products. Under certain oxidation conditions, the 1,10-isomer (I) can be converted to 2,2'-bipyridine-3,3'-dicarboxylic acid (VIII) or 1*H*-cyclopenta[2,1-*b*:3,4-*b'*]dipyridine-2-5-dione (IX) as reported by Baxter et al [22].

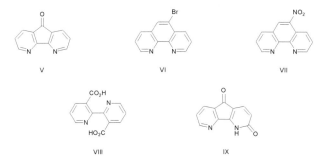

FIGURE 2 Potential impurities in 1,10-phenanthroline-5,6-dione (I).

Impurity profiling is an important issue in pharmaceutical analysis, particularly during product development and quality control. The standard requirements of such an impurity method [23] are that all likely synthetic and degradative impurities are resolved from each other and the main drug, and [24] that the impurities can be monitored at the 0.1% (w/w) level or below. Guidance for controlling impurity levels in drug substances has been developed by the International Conference on Harmonisation (ICH) [25]. Normally, synthetic impurities are discovered during routine HPLC analysis of the drug substance [26]. An impurity profile of a synthetic drug may require the use of complementary chromatographic methods such as HPLC/diode array UV and HPLC-MS to permit the observation of non-UV absorbing synthetic impurities.

The combination of MS and photodiode array detection (PDA) provides a powerful tool for controlling quality in drug synthesis and for the identification of impurities. The PDA detector has the capability to acquire and store a great amount of spectral data from the UV-absorbing compounds in chromatograms, thereby making

93

possible both spectral identification and individual analysis of the peak homogeneity/purity of each chromatographic peak.

GLC [27] has been used previously for the determination of phanquone (III) in biological material. Here, in this article, we describe a validated HPLC method for the assay of the quinone (I) and the application of HPLC-MS for identification of impurities in this compound.

2. EXPERIMENTAL

2.1 Chemicals and Reagents

All chemicals and reagents were of the highest purity. HPLC-grade methanol and triethylamine were obtained from Merck (Darmstadt, Germany). Water was purified using a Milli-Q system (Millipore, Watford, UK). 1,10-phenanthroline-5,6-dione (I) was prepared by the oxidation of 1,10-phenanthroline itself [16] and its cobalt complex [13]. Anhydrous 1,10-phenanthroline (I) was supplied by Lancaster Synthesis (Lancaster, UK). 4,5-Diazafluoren-9-one (V) was synthesized according to a literature method [11]. 5-Nitro-1,10-phenanthroline (VII) was obtained from Aldrich (Gillingham, UK).

2.2 HPLC-MS Instrumentation

HPLC-MS analysis was performed using a Waters ZQ2000 single quadrupole mass spectrometer, a Waters Alliance 2690 Separations Module to a 996 Waters PDA detector system (Waters, Elstree, UK). A stainless steel C18 column (250 × 2.0 mm i.d., 5 μm particle size) was used. A 2µl aliquot of a 0.6mg/ml solution of 1,10-phenanthroline-5,6-dione (I) was introduced by the HPLC system into the mass spectrometer via flow injection in a mobile phase of methanol:water (50:50, v/v) at 0.8 ml/min. The sample was ionised by positive-ion electrospray ionisation (ESI) probe in the positive ion mode using atmospheric pressure chemical ionisation (APCI) under the following source conditions: source temperature, 100°C; capillary potential, 3.0 kV; sampling cone potential, 30V. Mass spectra were obtained over the scan range 100-650 Da at a rate of 1 scan per second with 15.0 resolution and wavelength range 210 to 400 nm.

2.3 HPLC Instrumentation

The HPLC system consisted of a Waters Alliance 2690 Separations Module to a 996 Waters PDA Detector. Chromatographic separation was achieved isocratically in reversed-phase columns of the following type: Luna column C18 (150 × 4.60 mm i.d.) from Phenomenex UK (Macclesfield, UK), packed with silica, lowest silanol activity (100 Å, 5µm particle size, surface area 400 m^2/g, 17.5% carbon loading) was used. The mobile phase comprised methanol:water (50:50, v/v) containing 0.1% triethylamine and was filtered through a 0.45µm paper filter (Type HUHP, Millipore, Watford, UK) before use. Degassing of the mobile phase was carried out continuously with an on-line degasser. Flow rate was 0.8 ml/min. The system was equilibrated for approximately 30 min. The PDA detector was set at 254 nm. The column temperature was held at 40°C. The chromatographic run time was 10 min. The control of the HPLC system and data collection was by a Compaq computer equipped with Waters Millenium[32] software (version 3.20).

2.4 Preparation of the Standard and Sample Solutions

All sample and standard solutions at 0.6 mg/ml were prepared by dissolving approximately 60 mg of 1,10-phenanthroline-5,6-dione (I) in 100 ml HPLC grade methanol.

2.5 Linearity Assessment

Linearity experiments were performed by preparing dione (I) standard solutions in the range 0.04 – 1.60 mg/ml in methanol and injected in duplicate. Linear regression analysis was carried out on the standard curve generated by plotting the concentration of dione (I) versus peak area response.

2.6 Calculations and Other HPLC Determinations

Identity - the chromatographic profile of the sample preparation must show the same general profile (peak presence and relative intensities) as that of the appropriate standard.

Related substances – the quantity of each impurity peak (the known impurities designated as peaks (a) and (b) plus any other impurity peak) is calculated as an area percent versus the total area of all peaks in the chromatogram. Each impurity peak, plus the total area percent of all impurity peaks, must fall within the requirements of the specifications (Table 1).

Assay – the area percent of the main dione (I) peak, designated as peak (c), is determined in a similar manner to the impurities. The area percent must fall within the specifications (Table 1).

Purity – the purity of the dione (I) drug substance is calculated in relation to the reference standard using the area of the main peak (3):

Percent purity = $\dfrac{\text{mg/100ml of reference std.} \times \text{total area of peak 3 from sample}}{\text{area of peak 3 from reference standard} \times \text{mg/100ml of sample}}$

TABLE 1 Limits for the chromatographic purity of 1,10-phenanthroline-5,6-dione (I) and its impurities at 254 nm.

	Related substances		Assay
Peak assignment	1(a)*	2(b)*	3(c)*
Retention time (approximate, minutes)	1.4	2.1	2.5
Limit as % total area	≤ 2.0%	≤ 2.0%	
	a+b+all other impurities No other peak area can exceed 2.0%	≤ 4.0%	c ≥ 96.0%

*a = impurity 1; b = impurity 2; c = 1,10-phenanthroline-5,6-dione (I)

2.7 Forced Degradation Studies

Solutions of dione (I) were exposed to 50°C for one hour, UV light using a Mineralight UVGL-58 light for 24 hours, acid (1M HCl) for 24 hours and base (1M NaOH) for 4 hours.

2.8 Precision

The precision of the method was investigated by performing ten determinations of the same batch of dione (I) at 100% of the test concentration by only one operator. The repeatability (within-run precision) was evaluated by only one analyst within one day, whereas reproducibility (between-run precision) was evaluated for two different days.

2.9 Stability of Analytical Solutions

Test solutions of three batches of dione (I) were prepared using the conditions cited in section 2.4. They were chromatographed at the beginning and after 24 hours. The stability of dione (I) and the mobile phase were calculated by comparing area response and area percent of two standards at 0.6 mg/ml over time.

3. RESULTS AND DISCUSSION

3.1 Identification of Impurities by HPLC-MS

Good separation of 1,10-phenanthroline-5,6-dione (I) from its synthetic impurities was achieved by HPLC as demonstrated by the chromatogram displayed in Figure 3.

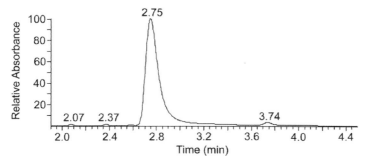

FIGURE 3 HPLC chromatogram of crude 1,10-phenanthroline-5,6-dione (I) displaying its impurities. This material was produced via the one-step oxidation method and is not crystallised.

The chromatographic run yielded four major peaks that are detailed by area percent in Table 2. Here, the data is from the analysis of a sample of crude (I) derived from the one-step oxidation of 1,10-phenanthroline.2d

TABLE 2 Chromatographic results for 1,10-phenanthroline-5,6-dione (I) and its impurities

Compounds	Retention times (min)	Peak area (μVs)	Peak area (%)
Impurity 1	2.07	25404	0.64
Impurity 2	2.37	23658	0.59
1,10-phenanthroline-5,6-dione (I)	2.75	3725059	93.21
Impurity 3	3.74	107954	2.70

The first two peaks at (2.07 and 2.37 min) are minor impurities while the third (3.74 min) is the major impurity. The peak at 2.75 min is identified as that due to 1,10-phenanthroline-5,6-dione (I) since its UV spectrum matches that of a known sample of (I) as shown in Figure 4.

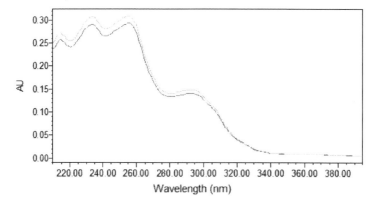

FIGURE 4 PDA UV match spectra of the middle of the peak corresponding to the retention time of the main component of 1,10-phenanthroline-5,6-dione (I) and a reference sample. See section 2.3 for HPLC conditions.

It was not possible to identify the chemical structure of the impurities by HPLC alone. For this purpose, HPLC-MS was selected to obtain structural information. The full scan HPLC-MS spectra of dione (I) and its impurities were measured in the mass range m/z 100-650. Dione (I) displayed a single peak at 2.75 min with a corresponding protonated molecular ion of mass m/z 211.2. Impurity peak 3 at 3.74 min was assigned to 4,5-diazafluoren-9-one (V) on the basis of both MS (Figure 5) and HPLC (Figure 6) data. Again, a protonated molecular ion of mass m/z 183.3 was observed in the mass spectrum.

The chemical structure of impurities 1 and 2 could not be identified from the HPLC and HPLC-MS data. However, the ions at m/z 183.3

and 241.2 for the two impurities (2.07 and 2.37 min, Fig. 5) do not correspond with the other suspected impurities such as 1,10-phenanthroline, 5-bromo-1,0-phenanthroline (VI), 5-nitro-1,10-phenanthroline (VII), 2,2'-bipyridine-3,3'-dicarboxylic acid (VIII) or 1*H*-cyclopenta[2,1-*b*:3,4-*b'*]dipyridine-2-5-dione (IX).

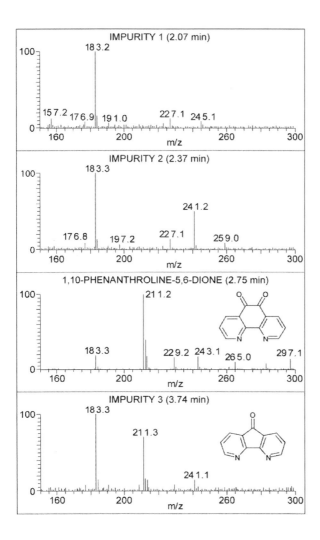

FIGURE 5 Mass spectra (x-axis: relative abundance) of 1,10-phenanthroline-5,6-dione and its impurities 1-3. Conditions: ESI+, source temperature 100°C, capillary voltage 3.0 kV, cone voltage 30V.

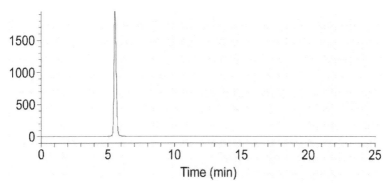

FIGURE 6 Separation of 4,5-diazafluoren-9-one (V), (impurity 3). Conditions: isocratic elution with methanol/water (50/50, v/v) with 0.1% of triethylamine, flow rate 0.8 mL/min, Luna column C_{18} (150 x 4.60 mm, 5-μm particle size), temperature 40°C, injection volume 2μL, UV detection 254 nm.

3.2 Syntheses

1,10-Phenanthroline-5,6-dione (I) was prepared according to the two methods in Scheme 1. Direct oxidation has the advantage in being a rapid one-step route from the readily available starting material 1,10-phenanthroline. As such, this method has proved very popular in the literature. In contrast, the cobalt complexation route is longer with two additional steps involving initial formation of the cobalt-phenanthroline compound and then final decomplexation of the product (I) using EDTA. Overall yields are moderate at 35 - 40% based on 1,10-phenanthroline. However, the direct oxidation step proceeds more cleanly and the cobalt-phenanthroline dione complex is isolated directly by precipitation with perchlorate ion. There is no requirement to neutralise the strongly acidic reaction mixture with the consequent risk of forming 4,5-diazafluoren-9-one (V). As such, the three-step route to (I) affords a purer product where (V) is completely absent (Figure 7). The crude dione (I) obtained from the one-step method can be purified by crystallisation from hot methanol solution which is filtered prior to cooling. The resulting crystallised material contains a reduced level of (V) as detected by HPLC. The unidentified impurities 1 and 2 appeared in samples of dione (I) produced by both synthetic methods.

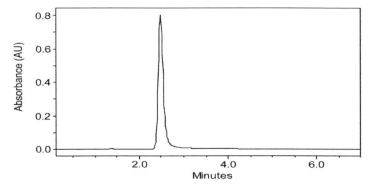

FIGURE 7 HPLC chromatogram of 1,10-phenanthroline-5,6-dione (I) and its potential impurities. This material was produced via the three-step route and is crystallised.

3.3 HPLC Method Development

Efficient chromatography and high sensitivity was achieved by using methanol-water as the mobile phase with varying detection wavelengths, based on the response of the active. However, the main peak tailed badly on some C18 columns with these mobile phases. Addition of triethylamine minimised the tailing. The amount of organic modifier was adjusted so that the assay run time could be reduced for faster analysis of samples. A chromatogram illustrating the separation of 1,10-phenanthroline-5,6-dione (I) and the two potential impurities is illustrated in Figure 7 confirming specificity with respect to dione (I).

The remaining chromatographic conditions listed in HPLC experimental section were chosen for the following reasons: the lower flow rate of 0.8 ml/min was chosen because of the potential problems associated with elevated back pressures. The PDA UV detector was set at 254 nm, λ_{max} for dione (I).

Column temperature was held at 40°C although separations at 30°C and 35°C indicated that slight variations in temperature did not have a significant effect on retention, resolution or peak shape. The injection volume of 2 µl and sample concentration of 0.6 mg 1,10-phenanthroline-5,6-dione/ml in methanol were chosen to simplify sample preparation (further dilution is not needed). This concentration allows both assay (of the main component) and purity evaluation (of trace impurities). The peak for dione (I) is well within the linear range for UV detection, and trace components are readily detectable.

A system suitability test was developed for the routine application of the assay method. Prior to each analysis, the chromatographic system must satisfy suitability test requirements (resolution and repeatability). Peak-to-peak resolution, between each peak measured on a reference solution, must be above 1.0. The %RSD of the response factor (area:mass ratio) for dione (I) sample peaks was determined from six replicate injections of the reference solutions and is required to be less than 2.0%.

System suitability testing was performed to determine the accuracy and precision of the system from six replicate injections of a solution containing 0.6 mg 1,10-phenanthroline-5,6-dione/ml. All peaks were well resolved and the precision of injections for all peaks was acceptable. The %RSD of the peak area responses was measured, giving an average of 0.62 (N = 6). The tailing factor (T) for each dione (I) peak was 1.61 and the theoretical plate number (N) was 3035.45. The resolutions between each peak were > 1.2 and the % RSD of retention time (RT) was $\leq$ 2% for six injections.

Selectivity was also studied over extended time using several columns and many different batches of mobile phase. Relative retention time (RT) ranges (RT of peak of interest/RT of 1,10-phenanthroline-5,6-dione) were as follows: a = 0.45; b = 1.20; c = 1.60. These data indicate that the RT windows for each impurity are unique and do not overlap. For rugged separations such as this, impurity identification could be based on relative RT alone. It would not be necessary to inject an authentic standard of each impurity to confirm identification. Overall selectivity was established through determination of purity for each impurity peak using the PDA UV detector.

For the determination of method robustness within a laboratory, a number of chromatographic parameters were determined which included flow rate, temperature, mobile phase composition, and column from different lots. In all cases good separations were always achieved, indicating that the method remained selective for all components under the tested conditions.

3.4 Validation of the Method

3.4.1 Linearity and range

The linearity of the method should be tested in order to demonstrate a proportional relationship of response versus analyte concentration over the working range. It is usual practice to perform linearity experiments over a wide range of the nominal concentration of analyte [29]. This gives confidence that the response and concentration are proportional and consequently ensures that calculations can be performed using a single reference standard, rather than the equation of a calibration line. Good linearity of this method was seen in the concentration range 0.04 – 1.60 mg/ml. Table 3 displays an assessment of the linearity results.

TABLE 3 Linearity assessment of the HPLC method for the assay of 1,10-phenanthroline-5.6-dione (I) employing the analytical working standard dissolved in methanol.

Concentration of dione (I) (mg/ml)	Concentration as % of 0.6 mg/ml of dione (I)	Dione (I) peak area as mean of 3 injections (μVs) (day 1)	Dione (I) peak area as mean of 2 injections (μVs) (day 2)
0.0400	4.0	225761	241740
0.1000	10	577405	616281
0.2002	20	1126188	1401765
0.4004	40	2574365	3063222
1.0009	100	6561498	8541076
1.6016	160	10458276	13686533
Correlation coefficient	0.9998		0.9996
Intercept (%)	-90		-26
Equation for regression line y = 7E+06x - 90335			9E+06x - 263239

3.4.2 Accuracy / recovery studies

Recovery studies may be performed in a variety of ways depending on the composition and properties of the sample matrix. In the present study, a number of different solutions were prepared with a known added amount of dione (I) and injected in triplicate. Percent recoveries of response factor (area/concentration) were calculated as shown in Table 4.

TABLE 4   Recovery studies

Analyte level (percent of target)	Recovery (%) ($n = 3$)
10	100.00
20	100.00
40	100.00
100	99.79
160	99.76

3.4.3 Specificity / selectivity

The PDA three-dimensional chromatogram (Figure 8) demonstrates a good separation of the dione (I) peak (c) (RT = 2.5 min) and from the impurities (RT = 1.4 and 2.0 min) and of the impurities from each other. A wavelength of 254 nm was found to be the most effective compromise to accomplish the detection and quantification of the two impurities and the main dione (I) component in a single run.

The impurities and the dione (I) peaks are adequately resolved from each other; typical resolution values for the dione (I) peak are ≥ 1.5. This method demonstrates acceptable specificity.

Forced degradation studies were performed to evaluate the specificity of dione (I) and its impurities under four stress conditions (heat, UV light, acid, base). A summary of the stress results is shown in Table 5.

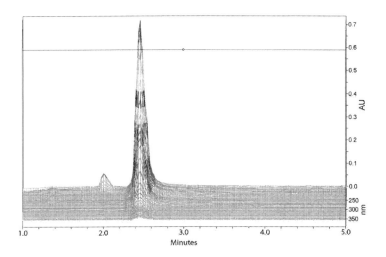

FIGURE 8 Three-dimensional chromatogram of the mixture of 1,10-phenanthroline-5,6-dione (I) and two of its impurities.

TABLE 5 Assay % of 1,10-phenanthroline-5.6-dione (I) under stress conditions

Stress conditions	Sample treatment	Retention time (min)	Assay %	Peak area (μVs)
Reference	Fresh solution	2.35	99.91	8582691
Acid	1N HCl for 24 hour	2.30	99.94	8753142
Base	1N NaOH for 4 hour	2.35	99.50	8697192
Heat	50°C for 1 hour	2.35	99.80	8630907
Light	UV Light for 24 hour	2.30	99.25	7189035

It is evident from Figure 7 that the method has been able to separate the peaks due to the degraded products from that of the dione (I). This was further confirmed by peak purity analysis on a PDA UV detector.

3.4.4 Precision

Percent recoveries were obtained from within- and between-run precision studies (Table 6). The %RSD values for the within run and between run precision studies are < 1%, thereby confirming that the method has acceptable precision.

TABLE 6 Comparison of within and between-run precision studies for 1,10-phenanthroline-5,6-dione (I).

Injection #	%Recovery day 1	%Recovery day 2
1	99.93	100.00
2	99.88	98.62
3	99.87	99.95
4	99.86	99.84
5	99.84	99.94
6	99.83	99.83
7	99.83	99.83
8	99.81	99.95
9	99.80	99.95
10	99.79	99.95
Average	99.84	99.79
%RSD	0.04	0.44
Total average	99.82	
Total %RSD	0.035	

3.4.5 Limit of detection and quantification

The limit of detection (LOD) and limit of quantification (LOQ) tests for the procedure were performed on samples containing very low concentrations of analyte. LOD is defined as the lowest amount of analyte that can be detected above baseline noise. Typically, this is three times the noise level. LOQ is defined as the lowest amount of analyte that can be quantified reproducibly above the baseline noise with a signal to noise ratio of 10. In this study the LOD was 2.0 µg/ml, and the LOQ was 200 µg/ml and RSD 0.45% ($n = 3$). An excellent match of the UV spectra of the impurities obtained at the limits of detection with those obtained at other concentrations (Figure 4, Table 7) was observed.

TABLE 7 Chromatographic parameters of 1,10-phenanthroline-5,6-dione (I) and its impurities and matching spectral results at LOD concentration levels.

Compound	Rt (min)	Purity (%)	Points across peak	Purity match angle[a] (°)	Purity threshold angle[b] (°)	λ_{max}
a	1.367	0.07	19	5.166	3.049	217.0
b	2.095	0.12	16	3.512	0.643	214.7
c	2.454	99.81	136	0.412	0.222	254.7

[a] Purity match angle = the spectral contrast between the two peaks

[b] Purity threshold angle = the spectral difference attributed to noise and solvent effects

3.4.6 Stability of analytical solutions

The stability of 1,10-phenanthroline-5,6-dione (I) solutions was investigated. The solutions were stable during the investigated 24 hours and the %RSD was in between 0.09 and 0.19% for retention times. Standard solutions stored in a capped volumetric flask on a laboratory bench under normal lighting conditions for 24 hours, were shown to be stable with no significant change in dione (I) concentration over this period (Table 8).

TABLE 8 Stability of 1,10-phenanthroline-5,6-dione (I) in solution ($n = 4$)

Time	Area %RSD	Height %RSD	% recovery	% of initial
0 hour	0.40	0.003	99.88	
24 hour	0.37	0.008	99.78	99.35

This is indicated (0.6% changes in area between $T = 0$ hour and $T = 24$ hours. Based on these data that show quantitative recovery through 24 hours, solutions of dione (I) can be assayed within 24 hours of preparation.

4. CONCLUSION

A reversed-phase HPLC method for the assay and purity evaluation of 1,10-phenanthroline-5,6-dione (I) obtained via two synthetic routes is described. The method has been demonstrated to be rugged and has been extensively validated. Application of the method in HPLC-MS mode resulted in the identification of 4,5-diazafluoren-9-one (V) as the major impurity in dione (I) prepared via the one-step oxidation of 1,10-phenanthroline. In contrast, the impurity (V) was absent from samples of dione (I) produced using a three-step cobalt complexation route from 1,10-phenanthroline. As such, this latter method is preferred for the synthesis of high purity samples of dione (I). Two minor impurities apparent in dione (I) synthesised via both routes remained unidentified.

REFERENCES

1. F. Kradolfer and L. Neipp, *Antibiot. Chemother.*, 8, (1958) 297.
2. T. S. Eckert, T. C. Bruice, J. A. Gainor and S. M. Weinreb, *Proc. Natl. Acad. Sci. USA*, 79, (1982) 2533.

3. T. S. Eckert and T. C. Bruice, *J. Am. Chem. Soc.*, 105, (1983) 4431.

4. S. Itoh, M. Kinugawa, N. Mita and Y. Ohshiro, *J. Chem. Soc., Chem. Commun.*, (1989) 694.

5. G. Hilt and E. Steckhan, *J. Chem. Soc., Chem. Commun.*, (1993) 1706.

6. S. Itoh, H. Fukushima, M. Komatsu and Y. Ohshiro, *Chem. Lett.*, (1992) 1583.

7. *Euro. Pat. Appl.*, 1,023,455 (2 Aug 2000); WO 9,919,507 (22 Apr 1999).

8. H. A. Byrne, K. L. Tieszen, S. Hollis, T. L. Dornan and J. P. New, *Diabetes Care*, 23, (2000) 500.

9. G. F. Smith and F. Cagle, *J. Org. Chem.*, 12, (1947) 781.

10. F. Linker and R. L. Evans, *J. Am. Chem. Soc.*, 68, (1946) 403.

11. J. Druey and P. Schmidt, *Helv. Chim. Acta*, 33, (1950) 1080.

12. *Brit. Pat.*, 688,802 (11 March 1953).

13. S. Imor, R. J. Morgan, S. Wang, O. Morgan and A. D. Baker, *Synth. Commun.*, 26, (1996) 2197.

14. J. E. Dickeson and L. A. Summers, *Aust. J. Chem.*, 23, (1970) 1023.

15. R. D. Gillard, R. E. E. Hill and R. Maskill, *J. Chem. Soc. A*, (1970) 1447.

16. M. Yamada, Y. Tanaka, Y. Yosimoto, S. Kuroda and I. Shimao, *Bull. Chem. Soc. Jpn.*, 65, (1992) 1006.

17. C. Hiort, P. Lincoln and B. Nordén, *J. Am. Chem. Soc.*, 115, (1993) 3448.

18. F. Calderazzo, F. Marchetti, G. Pampaloni and V. Passarelli, *J. Chem. Soc., Dalton Trans.*, (1999) 4389.

19. G. E. Inglett and G. F. Smith, *J. Am. Chem. Soc.*, 72, (1950) 842.

20. J. Mlochowski, *Rocz. Chem.*, 48, (1974) 2145.

21. J. Mlochowski and Z. Skrowaczewska, *Rocz. Chem.*, 47, (1973) 2255.

22. P. N. W. Baxter, J. A. Connor, J. D. Wallis, D. C. Povey and A. K. Powell, *J. Chem. Soc. Perkin Trans. 1*, (1992) 1601.

23. D. Zhong, H. Blume, *Pharmazie*, 49, (1994) 736.

24. J. Martens, *J. Chromatogr. B*, 673, (1995) 183.

25. International Conference on Harmonisation: Guidelines availability: Impurities in new drug substances: Notice, Fed. Reg., 61, (1996) 371-376.

26. S.V. Prabhu, *Talanta*, 40, (1993) 989.

27. P. H. Degen, S. Brechbühler, J. Schäublin and W. Riess, *J. Chromatogr. A*, 118, (1976) 363.

28. Guidelines for Collaborative Study Procedure to Validate Characteristics of a Method of Analysis, *J. Assoc. Off. Anal. Chem.*, 72 (1989) 694-704.

6 METHOD DEVELOPMENT AND VALIDATION FOR THE DETERMINATION OF 4,5-DIAZAFLUOREN-9-ONE USING HPLC AND IDENTIFICATION OF ITS IMPURITIES BY LC-MS

ABSTRACT

A new high-performance liquid chromatography-mass spectrometry (HPLC-MS) method was developed and validated for the identification and determination of novel 4,5-diazafluoren-9-one compound. The method employed a Waters XTerra RP-18 column (150 mm × 4.6 mm, id. 5 μm) with a mobile phase comprised a (50:50, v/v) mixture of deionised water containing 0.2% acetic acid (solvent A) and methanol (solvent B) at a flow rate of 1 mL/min, at 35°C. The detection was performed with photodiode-array (PDA) set at 210-400 nm and single quadropole mass spectrometer with electrospray ionization (ESI) positive ion mode. The chromatographic separation was achieved in less than 3 min. The linearity was established over the concentration range of 0.1-0.5 mg/mL (r^2 = 1.000). The mean RSD values for intra- and inter-day precision studies were <2%. The recovery of 4,5-diazafluoren-9-one ranged between 99.84 and 99.97%. The limits of detection and quantitation were determined to be 0.58 and 0.1 mg/mL, respectively.

Keywords: 4,5-diazafluoren-9-one, HPLC, HPLC-MS, Method development, Method validation,

1. INTRODUCTION

4,5-Diazafluoren-9-one (Figure 1b) is a useful co-product of the 1,10-phenanthroline (**a**). It can be synthesized by oxidation of 1,10-phenanthroline with alkaline potassium permanganate (KM_nO_4) in a potassium hydroxide (KOH) solution [1-3]. 4,5-Diazafluoren-9-one ($C_{11}H_6N_2O$) has hitherto been a difficulty accessible compound, its only synthesis being from the action of aqueous alkali on 1,10-phenanthroline,5-6-quinone by way of a benzilic acid type rearrangement analogous to the formation of fluorenone from phenanthrenequinone [4]. Various analytical procedures for characterization of 4,5-diazafluoren-9-one, including elemental analysis, UV-Vis, IR [5], NMR spectroscopy [6,7], X-ray crystallography and FTIR spectrophotometry [8] have been reported in the literature. However, HPLC or LC-MS method has not been reported in the literature.

LC is a universal separation technique that is capable of separating both volatile and non volatiles without the need for derivatization. Combined chromatographic and spectrometric techniques and in particular LC-MS have been contributing to the progress in life sciences in general [9]. In recent years, LC-MS has been extensively adopted in the field of biology, biochemistry, structural biology and biomedicine [10-18]. It is, therefore, promising and useful to develop an LC-MS method for the assay of 4,5-diazafluoren-9-one compound.

In this work for the first time, I report a simple, rapid, specific and sensitive LC-MS (with ESI+ ion mode) novel assay method for the separation, identification and determination of 4,5-diazafluoren-9-one compound. There are various types of ionization sources that can be used as the interface between the LC eluent and the mass spectrometer. The two most common sources are electrospray ionization (ESI) and atmospheric pressure chemical ionization (APCI), both of these source types are now standard equipment on mass spectrometers that are used for LC-MS apparatus. For both ESI and APCI, the ionization occurs at atmospheric pressure, so these sources are often referred to as atmospheric pressure ionisation (API) sources. For both ESI and APCI, some combination of high voltage and heat is used to provide the ionization that is needed to produce the ions that are assayed by the MS system. In ESI, the high voltage field (3-5kv) produces nebulization of the column effluent resulting in charged droplets that are focused toward the mass analyzer. These droplets get smaller as they approach the entrance to the mass analyzer: as the droplets get smaller, individual ions emerge in a process referred to as 'ion evaporation' – these ions are then separated by the MS system [19]. In APCI, heat is used to vaporize the column eluent and then a corona discharge is used to ionize solvent molecules, which then produce the analyte ion via chemical ionization mechanism [19].

Finally, the developed analytical method was validated to assess the validity of research data means determining whether the method used during the study can be trusted to provide a genuine, account of the intervention being evaluated. As a best practice [20-22], in the subsequent investigation, the new HPLC-MS method was validated [23] using pre-approved protocol and validated HPLC-MS system.

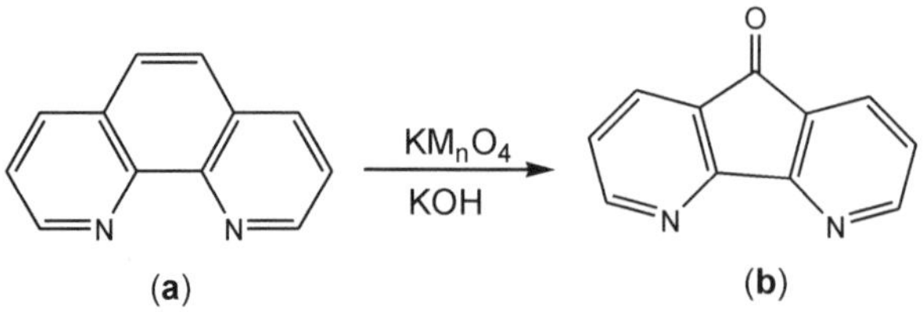

(a) $\xrightarrow[\text{KOH}]{\text{KM}_n\text{O}_4}$ (b)

FIGURE 1 Synthesis of 4,5-diazafluoren-9-one.

2. EXPERIMENTAL

2.1 Chemicals and Reagents

Methanol (HPLC-grade) and acetic acid were obtained from Merck (Darmstadt, Germany). 4,5-Diazafluoren-9-one reference standard and sample (98% pure) were purchased from Aldrich (St. Louis, Missouri, USA). Distilled water was de-ionised by using a Milli-Q system (Millipore, Bedford, MA).

2.2 Liquid Chromatography Conditions

A Waters (Elstree, UK) Alliance 2690 Separations Module LC system, consisting of a vacuum degasser, a quaternary solvent pump, an autosampler with a column oven and a photodiode-array

(PDA) detector, all controlled by a Empower software, was used. A reversed-phase XTerra RP-18 (Waters, Elstree, UK) column (150 × 4.6 mm, particle size 5 μm) was used for separation. The mobile phase comprised a (50:50, v/v) mixture of deionised water containing 0.2% acetic acid (solvent A) and methanol (solvent B). The flow rate was 1 mL/min, the injection volume was 4 μL and the temperature was set at 35°C. Chromatograms were recorded at 243 nm using UV detector. The second LC system used was PerkinElmer (Norwalk, CT, USA) equipped with a module series 200 UV detector, series 200 LC pump, series 200 autosampler and series 200 peltier LC column oven. The data were acquired via PE TotalChrom Workstation data acquisition software (version 6.3.1) using PE Nelson series 600 LINK interfaces.

2.3   Mass Spectrometry Conditions

A Waters (Elstree, UK) Micromass ZQ2000 single quadropole mass spectrometer (MS) with an ESI and APCI interfaces, all controlled by Empower software (version 1.0) was used. Operating conditions of the ESI interface in positive mode, which were obtained after making flow injection analysis (FIA) tests of the MS parameters, are reported in Table 1. Full-scan LC-MS spectra were obtained by scanning from *m/z* 100 to 500 da.

2.4 Preparation of the Standard and Sample Solutions

For each sample, approximately 30 mg of 4,5-diazafluoren-9-one was weighed into separate 100 mL volumetric flasks followed by the addition of 70 mL of HPLC-grade methanol. The resulting mixture was sonicated for about 5 minutes to aid dissolution of 4,5-diazafluoren-9-one. Volume made up to 100 mL with HPLC-grade methanol and was mixed well manually.

2.5 Linearity Assessment

Linearity experiments were performed by preparing 4,5-diazafluoren-9-one standard solutions in the range 0.1 – 0.5 mg/mL in methanol and injected in triplicate. Linear regression analysis was carried out on the standard curve generated by plotting the concentration of 4,5-diazafluoren-9-one versus peak area response.

3. RESULTS AND DISCUSSION

3.1 Liquid Chromatography Analysis

Prior to the coupling with the mass spectrometer, an optimization of the liquid chromatographic separation was carried out using photodiode array UV detector.

Acetic acid in water, ammonium formate and ammonium acetate buffers were studied, selecting the first one because less analysis time and better separation were obtained with the addition of methanol. Initially, three analytical columns were tried in order to reach acceptable specificity and selectivity. Initially, exploited Luna C$_{18}$ (150 × 4.6 mm, 5-µm) and HyperClone (ODS) C$_{18}$ (250 × 4.6 mm, 5-µm) phase's columns from Phenomenex (Macclesfield, UK). The Luna column gave poor separation with long tailing at fronting of analyte peak with retention time 2.96 min. The HyperClone column gave good peak shape but retention time was 4.46 (min) for the 4,5-diazafluoren-9-one peak. Shift to XTerra RP-18 (150 × 4.6 mm, 5-µm) column (Waters, Elstree, UK) produces peak with superior band shape and column efficiency with shorter retention time (2.65 min) under the same conditions (Figure 2).

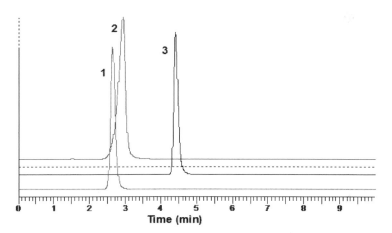

FIGURE 2 HPLC chromatograms obtained from different C$_{18}$ columns: 1) Xterra, 2) Luna, 3) HyperClone.

Some tests were made varying the temperature between 25°C and 55°C at 10°C steps, to study the influence of this parameter. The results showed that the variation of the temperature at 25-45°C did not affect significantly any of the chromatographic parameters and only increased the retention time of the analyte (Figure 3). At higher temperature (55°C), worsened the peak shape and selectivity (Figure 4), so 35°C was chosen as work temperature.

The choice of wavelength is essential to accomplish a sensitive and a selective chromatographic assay. The optimal wavelength for 4,5-diazafluoren-9-one detection was established using the scan range of 190 to 400 nm, It was shown that 243 nm were the optimal wavelength to maximize the signal (Figure 5).

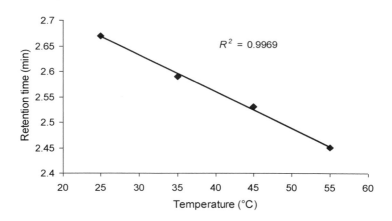

FIGURE 3 Effect of HPLC column temperature on the retention times.

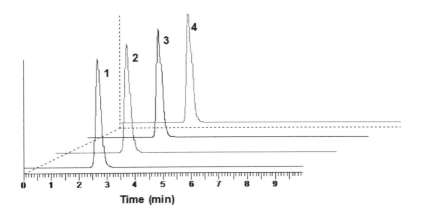

FIGURE 4 LC chromatograms of 4,5-diazafluoren-9-one obtained at different temperatures: (1) 25°C, RT, 2.67, (2) 35°C, RT, 2.57 (3) 45°C, RT, 2.53 and (4) 55°C, RT, 2.45 (min).

To evaluate the quantitative nature of the analytical method, a series of samples with different amounts of 4,5-diazafluoren-9-one were run to investigate the best assay concentration. Using an XTerra RP-18 column, best concentration was assessed by injecting six-reference standard of 4,5-diazafluoren-9-one in the range of 0.01 to 1.0 mg/mL. The integrated peak areas were plotted versus amount injected. The calibration curve was found to be linear from concentration range 0.1 to 0.5 mg/mL with a correlation coefficient of 0.9999. On the bases of these data, the middle concentration of the linearity was chosen as a best concentration (0.3 mg/mL) for the assay.

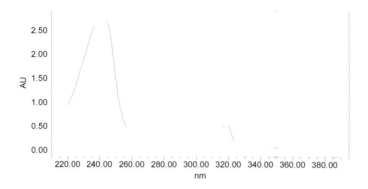

FIGURE 5 Photodiode-array UV spectra of the middle of the peak corresponding to the retention times of main compound of 4,5-diazafluoren-9-one.

The system suitability test was established from six replicate injections of a solution containing 0.3 mg 4,5-diazafluoren-9-one /mL. The percent relative standard deviation (RSD) of the retention time (min) and peak area were found to be less than 0.50%. The USP tailing factor, T_f, was 0.76, and column efficiency, N, was 6012 for 4,5-diazafluoren-9-one.

Robustness testing was performed during method development phase to optimize final LC conditions. An LC method must prove to be able to remain unaffected by small, but deliberate variations in method parameters, thus showing its own reliability during normal usage. It is advisable to simultaneously study the possible variations of method parameters in an interval chosen symmetrically around the optimised conditions. This interval represents the variations expected during method transfer and routine use in quality control testing. In this case, the seven selected parameters were the same considered in the optimisation step. Their experimental domain is reported in Table 1. This showed that the method for determination of 4,5-diazafluoren-9-one was reproducible and robust.

3.2 Mass Spectrometry Analysis

The first MS experiments to select the optimum MS parameters and the appropriate ions were carried out by Flow Injection Analysis of 4,5-diazafluoren-9-one sample solutions, monitoring the MS intensity. Different ionization interfaces APCI and ESI modes (negative and positive ionization) were investigated. The chromatogram is generated from the ion abundances and mass spectra recorded in the positive-ion, ESI mode gave almost no loss of sensitivity compared to the APCI. The studied range for each ESI-MS parameter is shown in Table 2.

TABLE 1 Experimental domain of the factors during robustness testing

Factor	Experimental domain	Optimal value
Sample solvent	Mobile phase, methanol, water	Methanol
Analytical columns (Different lots)	C_{18}-C_{18}	C_{18}
Percent organic solvent	45-55	50
Flow rate (mL/min)	0.8-1.2	1.0
Injection volume (μL)	2-6	4
Column T (°C)	30-40	35
Wavelength (nm)	239-247	243

Best sensitivity was obtained for 4,5-diazafluoren-9-one compound using the conditions in Table 2. The full scan MS spectrum of 4,5-

diazafluoren-9-one was measured in the mass range m/z 100-500 da. The 4,5-diazafluoren-9-one displayed a single peak at 2.735 min, which corresponds to the molecular mass [M+H]$^+$ at m/z 183.04 (see Figure 6). In addition, Figure 7 shows three-dimensional display of the photodiode array absorbance data obtained by LC-PDA-MS for 4,5-diazafluoren-9-one. Peak at m/z 183.04 is for 4,5-diazafluoren-9-one. The first dimension is LC retention time, second is m/z and third is intensity.

TABLE 2 Results obtained from the FIA tests of the MS parameters

MS parameter	Studied range	Optimal value
Capillary voltage	2-4 kv	3.5 kv
Cone voltage	50-70 v	60 v
Extractor voltage	2-10 v	5 v
Source temperature	50-120 (°C)	80 (°C)
Cone temperature	10-30 (°C)	20 (°C)
Desolvation temperature	150-350 (°C)	200 (°C)
Desolvation gas flow	100-400 (L/Hr)	260 (L/Hr)
Cone gas flow	30-65 (L/h)	60 (L/h)

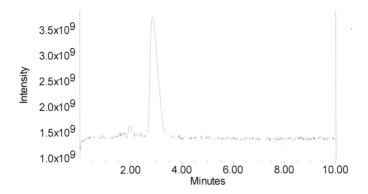

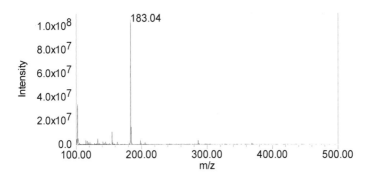

FIGURE 6 LC-MS chromatogram (Top), ESI (-) MS spectrum of 4,5-diazafluoren-9-one, (Bottom). The LC-MS retention time and the mass spectrum readily identify the peak eluting at 2.735 min as 4,5-diazafluoren-9-one.

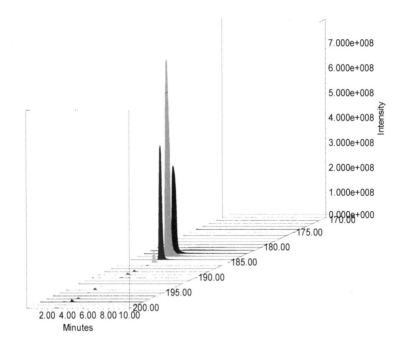

FIGURE 7 Three-dimensional display of the photodiode array absorbance data obtained by LC-PDA-MS for 4,5-diazafluoren-9-one. The first dimension is LC retention time, second is *m/z* and third is intensity.

3.3 Method Validation

Great care was taken in the method development phase, which resulted in the subsequent validation being straightforward. The validation parameters performed were linearity, range, precision (repeatability and intermediate precision), accuracy, specificity, and limit of detection and quantitation.

3.3.1   Linearity and range

Linearity was studied in the concentration range 0.1 to 0.5 mg/mL (n = 3; K = 5) and the following regression equation was found by plotting the peak area (y) versus the 4,5-diazafluoren-9-one concentration (x) expressed in mg/mL:

$$y = 32281938.00x + 194090.20 \ (r^2 = 1.000) \ (1)$$

The determination coefficient (r^2) obtained (Table 3) for the regression line demonstrates the excellent relationship between peak area and the concentration of 4,5-diazafluoren-9-one.

TABLE 3 Linearity assessment data for the assay of 4,5-diazafluoren-9-one

Concentration (mg/mL)	Concentration as % of analyte target	Mean area (μVs, n = 3)	$\pm$SD	RSD (%, n = 3)
0.1	10	3374350	15926.80	0.47
0.2	25	6642282	16319.09	0.25
0.3	75	9921267	10500.98	0.11
0.4	100	13237998	16625.44	0.13
0.5	150	16217461	12561.90	0.08

Correlation coefficient: r^2 = 1.000; Equation for regression line: y = 32281938.00x + 194090.20

3.3.2   Precision (repeatability and intermediate precision)

The precision of the chromatographic method, reported as percent RSD, was estimated by measuring repeatability (intra-day assay precision) on ten replicate injections at 100% test concentration (0.3 mg/mL). The RSD values for retention time (min) 0.04%, peak area 0.49% and peak height were 0.48%.

The intermediate precision (inter-day variation) was studied using two LC systems over two consecutive days at three different concentration levels (0.2, 0.3 and 0.4) that cover the assay range (80-120%). Three replicate injections were injected for each solution. The chromatograms obtained using Perkin Elmer (LC 1) and Waters (LC 2) are given in Figure 8 and 9, respectively. The RSD values for both instruments were ≤1.08% (Table 4) and illustrated the good precision of this analytical method.

TABLE 4 Intermediate precision study data for 4,5-diazafluoren-9-one

Concentration	LC 1, Day 1		LC 2, Day 2	
(mg/mL)	Mean area (μVs, n = 3)	RSD (%, n = 3)	Mean area (μVs, n = 3)	RSD (%, n = 3)
0.2	5320180	0.27	6745738	1.08
0.3	7736782	0.35	10060231	0.24
0.4	10208667	1.01	13351243	0.24

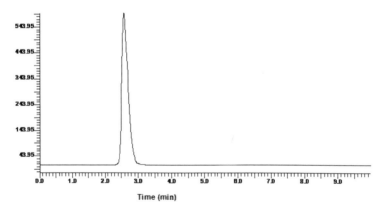

FIGURE 8 LC chromatogram of 4,5-diazafluoren-9-one eluted at 2.693 min obtained using Perkin Elmer LC system during inter-day variation study.

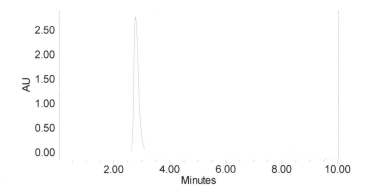

FIGURE 9 LC chromatogram of 4,5-diazafluoren-9-one eluted at 2.735 min obtained using Waters LC system during inter-day variation study.

3.3.3 Accuracy / recovery study

Recovery studies may be performed in a variety of ways depending on the composition and properties of the sample matrix. In the present study, a number of different solutions were prepared with known added amount of 4,5-diazafluoren-9-one and injected in triplicate. Percent recoveries of response factor (area/concentration) were calculated as can be seen in Table 5.

TABLE 5 Recovery studies data for 4,5-diazafluoren-9-one from samples with known concentration

Sample	Concentration as % of analyte target	Mean recovery (%, n = 3)	RSD (%, n = 3)
1	10	99.84	0.03
2	25	99.96	0.04
3	75	99.95	0.01
4	100	99.97	0.02
5	150	99.96	0.05

3.3.4 Specificity/forced degradation studies

The LC-PDA three-dimensional chromatogram (Figure 10) demonstrates a good separation of the 4,5-diazafluoren-9-one. The three-dimensional data consist of UV absorption spectra from 210 to 400 nm for each point along the chromatogram. This method demonstrates acceptable specificity.

The forced degradation studies were performed to evaluate the specificity of 4,5-diazafluoren-9-one under four stress conditions (heat, UV light, acid, base). Solutions of 4,5-diazafluoren-9-one were exposed to 50°C for 1 h, UV light using a Mineralight UVGL-58 light for 24 h, acid (1M hydrochloric acid) for 24 h and base (1M sodium hydroxide) for 4 h. A summary of the stress results is shown in Table 6. It is evident from Figure 11 that the method has been able to separate the peak due to the degraded products from that of the 4,5-diazafluoren-9-one.

TABLE 6 Force degradation studies data for 4,5-diazafluoren-9-one

Stress conditions	Sample treatment	RT (min)	Area (μVs)	Assay (%)
Reference	Fresh solution	2.69	6927585	99.86
Acid	1M HCl for 24 h	2.75	6862371	98.94
Base	1M NaOH for 4 h	2.64	6752936	98.51
Heat	50 °C for 1 h	2.66	6692625	99.92
Light	UV Light for 24 h	2.66	6957040	99.92

3.3.5 Limits of detection and quantitation

The limit of detection (LOD) and limit of quantitation (LOQ) of 4,5-diazafluoren-9-one were estimated from the intercept ($\bar{a}$) of the regression line and the corresponding residual standard deviation

($S_{y/x}$) (18). The responses at the LOD and LOQ were estimated by the following expressions respectively.

$f(\text{LOD}) = \bar{a} + 3S_{y/x}$ (2)

$f(\text{LOQ}) = \bar{a} + 10S_{y/x}$ (3)

Applying this method, LOD and LOQ for 4,5-diazafluoren-9-one were found to be 0.58 and 0.1 mg/mL, respectively.

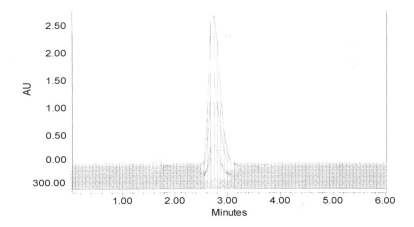

FIGURE 10 Three-dimensional display of the photodiode array absorbance data obtained by LC-PDA for a 4,5-diazafluoren-9-one.

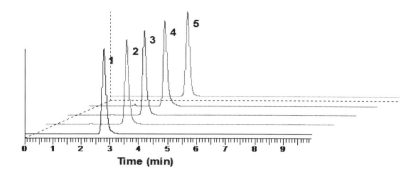

FIGURE 11 LC chromatograms obtained from forced degradation study: 1) fresh solution, 2) acid, 3) base, 4) heat at 50 °C, 5) UV light.

3.3.6 Stability of analytical solutions

The stability of 4,5-diazafluoren-9-one solutions was investigated. The solutions were stable during the investigated 48 hours and the RSD was in between 0.04 and 0.07% for retention times. Standard solutions stored in a capped volumetric flask on a laboratory bench under normal lighting conditions for 48 hours, were shown to be stable with no significant change in 4,5-diazafluoren-9-one concentration over this period (Table 7). This is indicated (0.5% changes in area between T = 0 hour and T = 48 hours. Based on these data that show quantitative recovery through 48 hours, solutions of 4,5-diazafluoren-9-one can be assayed within 48 hours of preparation.

TABLE 7 Stability of 4,5-diazafluoren-9-one in solution ($n = 3$)

Time (h)	Area (%RSD)	Height (%RSD)	Recovery (%)	% of initial
0	0.62	0.09	99.95	
48	0.48	0.12	99.88	99.93

4. CONCLUSION

A new, simple, sensitive and specific LC-MS method for the assay of 4,5-diazafluoren-9-one was developed. With the use of the Waters XTerra RP-18 column, 4,5-diazafluoren-9-one was well retained, with a good peak shape. A simple binary isocratic was used without adding buffer. The proposed method for its validity was successfully validated using validated (qualified) LC-MS system and results showed a good linearity, precision and accuracy. The quantification linear range was from 0.1 to 0.5 mg/mL with correlation coefficient (r^2) of 1.000. The mean RSD values for intra- and inter-day precision studies were <2% in each case. The developed method could be satisfactorily applied as a routine procedure to identify and quantify 4,5-diazafluoren-9-one.

REFERENCES

1. L.J. Henderson, F.R. Fronczek and W.R. Cherry. Selective perturbation of ligand field excited states in polypyridine ruthenium(II) complexes. *J. Am. Chem. Soc.* 106: 5876-5879 (1984).
2. T. Zihou, et al. Molecular assemblies of diazafluorenone Schiff-base amphiphiles. II. The vesicle and its molecular aggregation behaviour. *Molecular Engineering.* 3: 293-299 (1994).

3. O. Katsuhiko, et al. Synthesis and properties of 9,9'-diaryl-4,5-diazafluorenes, A new type of electron-transporting and hole-blocking material in EL device. *Chem. Lett.*, 33: 276-277 (2004).

4. G. M. Badger and J.W. Cook. Chemistry of Carbon Compounds, (Eds.) E.H. Rodd, Vol. IIIB, Elsevier, Amsterdam 1956, p.1446.

5. M.S. Deshpande and A.S. Kumbhar. Mixed-ligand complexes of ruthenium (II) incorporating a diazo ligand: synthesis, characterization and DNA binding. *J. Chem. Sci.*, 117: 153-159 (2005).

6. G. Zhou and I.I. Harruna. Synthesis of ligand monomers derived from 4,5-diazafluoren-9-one. *Tetrahedron Lett.* 44: 4617-4619 (2003).

7. U. Siemeling and I. Scheppelmann. Cyclopentadienone-like behaviour of fluorenone and 4,5-diazafluoren-9-one. *Organometallics*, 23: 626-628 (2004).

8. S. Menon and M.V. Rajasekharan. A channel-forming polyiodide network in [Cu(dafone)$_3$]I$_{12}$. A tris chelate of defone and a new planar structure for the I$_{12}^{2-}$ ion (defone = 2,5-diazafluoren-9-one). *Inorg. Chem.*, 36: 4983-4987 (1997).

9. E. Gelpi. Contributions of liquid chromatography–mass spectrometry to "highlights" of biomedical research. *J. Chromatogr. A*, 1000: 567-581 (2003).

10. A. Vera Francesca, L. Giovanna, P. Carlo, *et al.* Determination of protein phosphorylation sites by mass spectrometry : a novel electrospray-based method Rapid Commun. *Mass Spectrom.* 19: 3343-3348 (2005).

11. H.Y. Ji, H.W. Lee, H. Kim, H.K. Kim, *et al.* Liquid chromatography-mass spectrometric analysis of compound K, a ginseng saponin metabolite, in rat plasma. *Anal. Lett.* 37: 1307-1318 (2004).

12. R. Buchalla and T.H. Begley. Characterization of gamma-irradiated polyethylene terephthalate by liquid-chromatography–mass-spectrometry (LC–MS) with atmospheric-pressure chemical ionization (APCI). *Radiat. Phys. Chem.* 75: 129-137 (2006).

13. S.S. Singh and K. Sharma. Validation of LC-MS electrospray ionisation method for quantitation of haloperidol in human plasma and its application to bioequivalence study. *Anal. Chim. Acta* 551: 159-167 (2005).

14. K.P. Deventer, F. P. Van, W. Mikulcikova, T. Van, F.T. Delbeke. Quantitative analysis of androst-4-ene-3,6,17-trione and metabolites in human urine after the administration of a food supplement by liquid chromatography/ion trap-mass spectrometry. *J. Chromatogr. B*, 828: 21-26 (2005).

15. M. Xu, G. Wang, H. Xie, R. Wang, *et al*. Determination of schizandrin in rat plasma by high-performance liquid chromatography–mass spectrometry and its application in rat pharmacokinetic studies. *J. Chromatogr. B* 828: 55-61 (2005).

16. C.N. McEwen, R.G. Mckay, and B.S. Larsen. Analysis of Solids, Liquids, and Biological Tissues Using Solids Probe Introduction at Atmospheric Pressure on Commercial LC/MS Instruments. *Anal. Chem.* **77**: 7826-7831 (2005).

17. T. Qian, Z. Cai, R.N.S. Wong and Z-H. Jiang. Liquid chromatography/mass spectrometric analysis of rat samples for *in vivo* metabolism and pharmacokinetic studies of ginsenoside Rh2, Rapid Commun. *Mass Spectrom.*19: 3549-3554 (2005).

18. M. Liu, Y. Hashi, Y.Y. Song and J.M. Lin. Simultaneous determination of carbamate and organophosphorus pesticides in fruits and vegetables by liquid chromatography–mass spectrometry. *J. Chromatogr. A* 1097: 183-187 (2005).

19. W.M. Niessen. Progress in liquid chromatograpy-mass spectrometry instrumentation and its impact on high-throughput screening. *J. Chromatogr. A* 1000: 413-436 (2003).

20. G.A. Shabir, W.J. Lough, A. A. Shafique, T.K. Bradshaw. Evaluation and application of best practice in analytical method validation. *J. Liq. Chromatogr. Rel. Technol.*30: 311-333 (2007)>

21. G.A. Shabir. Validation of HPLC methods for pharmaceutical analysis: Understanding the differences and similarities between validation

requirements of the U.S. FDA, USP and ICH. *J. Chromatogr. A* 987: 57-66 (2003).

22. G.A. Shabir. Step-by-step analytical methods and protocol in the quality system compliance industry. *J. Validation Technol.* 10: 314-324 (2004).

23. International Conference on Harmonization (ICH), Q2(R1): Validation of analytical procedures: Text and Methodology, Nov 2005.

7 DEVELOPMENT AND VALIDATION OF A HPLC METHOD FOR THE DETERMINATION OF BENZO[*F*]QUINOLINE-5,6-DIONE AND IDENTIFICATION OF ITS IMPURITIES USING LC-MS

ABSTRACT

A new analytical method based on reversed-phase high-performance liquid chromatography (HPLC) was developed and validated for the assay of benzo[*f*]quinoline-5,6-dione and the determination of its synthetic impurities by employing the method in liquid chromatography-mass spectrometry (LC-MS) with electrospray ionization (ESI) and photodiode array (PDA) UV detection. Separation was performed on a Waters Xterra C_{18} (150 mm × 4.6 mm, 5 μm) column. UV detection was performed at 262 nm. The results showed that benzo[*f*]quinoline-5,6-dione is eluted as a spectrally pure peak resolved from its impurities. 1*H*-indeno[2,1-*b*]pyridine-2,9-dion and benzo[*f*]quinoline are identified as the main impurities. Proposed method was extensively validated and satisfactory results were obtained in terms of linearity (r^2 = 0.9997) and precision (RSD <0.61%) in all cases. The limits of detection and quantification were 2 μg/mL and 20 μg/mL, with a 0.88% RSD, respectively.

Keywords: HPLC, LC-MS, Method development, Validation, Benzo[*f*]quinoline-5,6-dione, Synthetic impurities

1. INTRODUCTION

Benzo[*f*]quinoline-5,6-dione **2** (4-azaphenanthrene-5,6-dione), is an important heterocyclic o-quinone, which is known historically to be active against protozoa, amoebae and bacteria [1]. Benzo[*f*]quinoline **1** and benzo[*f*]quinoline-5,6-dione **2** are starting materials for quinoline legends which are useful for the preparation of complexes of transition metals. Benzo[*f*]quinoline-5,6-dione was synthesized from benzo[*f*]quinoline by oxidation with iodine pentoxide (I_2O_5) in acetic acid (scheme 1). A number of potential impurities were expected from an examination of the literature [2] concerning the reactivity of the benzo[*f*]quinoline-5,6-dione (Figure 1). The purpose of this study was to develop a reversed-phase high-performance liquid chromatography (HPLC) assay method for benzo[*f*]quinoline-5,6-dione to asses its purity, while also applying HPLC-electrospray ionization (ESI)-mass spectrometry (MS) to the identification of any impurities. Impurity profiling is a major issue in pharmaceutical and biomedical analysis, particularly during product development phase and quality control.

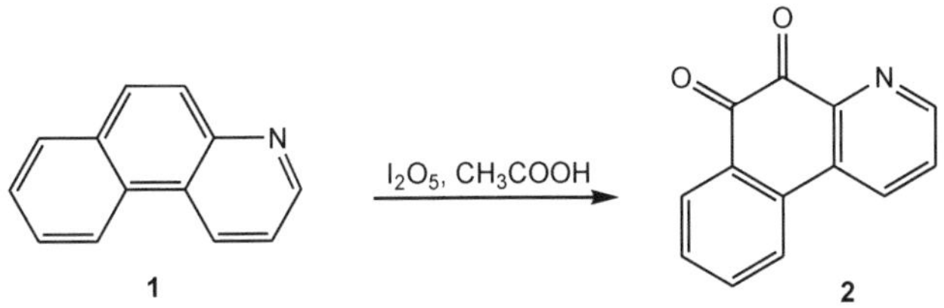

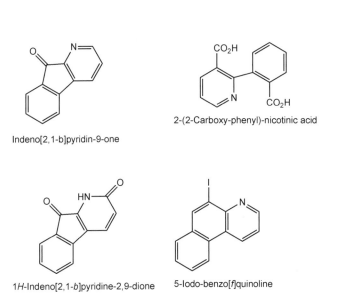

Scheme 1 Synthesis of benzo[f]quinoline-5,6-dione.

Indeno[2,1-b]pyridin-9-one

2-(2-Carboxy-phenyl)-nicotinic acid

1*H*-Indeno[2,1-*b*]pyridine-2,9-dione

5-Iodo-benzo[f]quinoline

FIGURE 1 Structures of potential impurities in benzo[f]quinoline-5,6-dione.

The standard requirements of such an impurity method [3] are that all likely synthetic and degradative impurities are resolved from each other and the main drug, and [4] that the impurities can be monitored at the 0.1% (w/w) level or below. Guideline for controlling impurity levels in drug substances has been developed [5].

Normally, synthetic impurities are discovered during routine HPLC analysis of the drug substances [6]. An impurity profile of a synthetic drug may require the use of complementary chromatographic methods such as HPLC/diode array UV and LC-MS to permit the observation of non-UV absorbing synthetic impurities. The combination of MS and PDA detection provides a powerful tool for controlling quality in drug synthesis and for the identification of impurities. The PDA detector has the capability to acquire and store a great amount of spectral data from the UV-absorbing compounds in chromatograms, thereby making possible both spectral identification and individual analysis of the peak homogeneity / purity of each chromatographic peak. ESI, an atmospheric ionization technique was used to generate gas-phase ions by spraying analyte solution at high voltage. ESI has become one of the most important and powerful ionization technique for MS because of its effectiveness in detecting large biomolecules [7] and ease-of-use for interfacing liquid-based separation techniques, such as LC [8,9] and capillary electrophoresis [10-12]. According to literature search, this is the first report on HPLC based separations and identifications of impurities of this benzo[*f*]quinoline-5,6-dione compound. Herein, my effort in developing and validating an HPLC method for this important novel compound and my further studies of its impurity profile by LC-MS are described. The aim of the validation work is to prove that the research data/results obtained in the analysis of benzo[*f*]quinoline-5,6-dione and its synthetic impurities are indubitable. This involves demonstration of aspects such as linearity, range, accuracy, precision, specificity, sensitivity, limit of detection and limit of quantitation. The results obtained in these evaluations show that the method could be used reliably in routine analysis of benzo[f]quinoline-5,6-dione.

2. EXPERIMENTAL

2.1 Chemicals and Reagents

Acetonitrile (HPLC-grade) was obtained from Merck (Darmstadt, Germany). Ammonium formate (Formic acid ammonium salt, $HCOONH_4$) was purchased from Sigma chemicals (St. Louis, MO, USA). Benzo[f]quinoline-5,6-dione was prepared by Charnwood Molecular (Loughborough, UK). Benzo[f]quinoline-5,6-dione standard was characterised in house (>92%, Pure). De-ionised distilled water was used throughout the experiment.

2.2 HPLC and LC-MS Conditions

A Waters 2695 Alliance Separations Module equipped with a 996 photodiode array detector (PDA) (Waters, Milford, MA, USA) was used. Separation was achieved on a 150 mm × 4.6 mm, 5 μm particle size Waters Xterra C_{18} column. The mobile phase consisted of 15 mM ammonium formate (pH 3) as solvent A and acetonitrile as solvent B, which were applied in the following gradient elution: initial 0-12 min, linear change from A-B (95:5, v/v) to A-B (65:35, v/v); 12-16 min, linear change to A-B (50:50, v/v); 16-25 min, linear change to A-B (40:60, v/v). The flow rate was 0.8 mL/min and sample injection volume was 10 μL. The temperature was held constant at

32°C. Ultraviolet (UV) spectra were recorded in the 210-400 nm range and the chromatograms were extracted at 262 nm. The data was collected and analyzed by Waters Empower software.

LC-MS analysis was performed using a Waters Micromass ZQ 2000 single quadrupole mass spectrometer, equipped with a Waters Alliance 2695 Separations Module and 996 PDA (Waters, Milford, MA, USA). Same LC conditions were used in LC-MS analysis as described above in HPLC conditions section. The sample was ionized using ESI in the positive ionization mode under the following source conditions: source temperature, 150°C; capillary potential, 3.0 kv; sampling cone potential, 30 v; cone gas flow, 60 L/h; extractor, 3.0 v; RF lens, 0.5 v; desolvation temperature, 350°C; low mass resolution 15; high mass resolution 15; ion energy, 0.3 v; and multiplier 430 v. Mass spectra were obtained over the scan ranges, 2-800 da, at a rate of 0.5 scan per second. A wavelength range 210 to 400 nm was used for the PDA detector. In ESI, a high electrical voltage charges the eluent as it emerges from a nebulizer, producing an aerosol of charged droplets. As the solvent evaporates, the droplets shrink, developing a charge dense enough to eject ions from their surfaces (ion evaporation). The mass analyzer then sorts the singly or multiply charged ions by mass-to-charge (m/z) ratio.

2.3 Sample and Standard Preparation

All sample and standard solutions at 0.20 mg/mL were prepared by dissolving approximately 20 mg of benzo[f]quinoline-5,6-dione in 100 mL of acetonitrile.

2.4 Calculations and Other HPLC Determinations

2.4.1 Related substances

The quantity of each known impurity peak plus any other impurity peaks was calculated as an area percent versus the total area of all peaks in the chromatogram.

2.4.2 Purity

The purity of the benzo[f]quinoline-5,6-dione drug substance is calculated in relation to the reference standard using the area of the main benzo[f]quinoline-5,6-dione peak at 9.187 min (Eq. 1).

$$\text{Peak purity} = \frac{\text{mg/100 mL of standard x total area of main analyte from sample}}{\text{Area of main analyte peak from standard} \times \text{mg/100 mL of sample}} \quad (\text{Eq. 1})$$

3. RESULTS AND DISCUSSION

3.1 HPLC Analysis

Prior to the coupling with the mass spectrometer, an optimization of the liquid chromatographic separation was carried out using photodiode array UV detector. Some important considerations have to be taken into account when a HPLC method is developed before it's coupling with mass spectrometry detection. For example, a volatile buffer of low conductivity (i.e., eclectic current below 50 μA) is required to avoid plugging of the dielectric capillary between the spray chamber and mass spectrometer, as well as to obtain a stable electrospray. Ammonium formate and ammonium acetate buffers were studied, selecting the first one because better peak shapes were obtained with the addition of acetonitrile. With this buffer, a systematic study of the pH effect was carried out between pH 2 and 4; finally pH 3 was chosen because at this value the best resolution between the main component and its impurities occurred. The optimized buffer concentration was15 mM as it was found to provide a good compromise among peak shape, and analysis time. Two analytical columns were tried in order to reach acceptable specificity and selectivity. Initially was exploited RP-8 columns but analyte retained on these columns. Shift to RP-18 columns, among which Xterra column proved to be superior to others, exhibited better separation with shorter retention time (Figure 2) probably due to its polymeric octadecylsilane of less silanol sites with wide pH range. The temperature of the analytical column was set to 32°C because lower as well as higher temperatures worsened the resolution among main component and its impurities. The choice of wavelength is essential to accomplish a sensitive and a selective chromatographic assay. The optimal wavelength for benzo[f]quinoline-5,6-dione detection was established using two UV absorbance scans over the range of 190 to 400 nm, one scan of the mobile phase, and the second of the analyte in the mobile phase. It was shown that 262 nm were the optimal wavelength to maximize the signal. To evaluate the quantitative nature of the analytical method, a series of samples with different amounts of benzo[f]quinoline-5,6-dione were run to investigate the best assay concentration. Using a C_{18} column, best concentration was assessed by injecting six reference standard of benzo[f]quinoline-5,6-dione in the range of 0.002 to 1.60 mg/mL. The integrated peak areas were plotted versus amount injected. The calibration curve was found to be linear from concentration range 0.02 to 1.40 mg/mL with a correlation coefficient of 0.9997. On the bases of these data, the best concentration (0.20 mg/mL) was chosen as a working concentration for the assay.

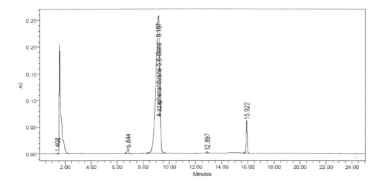

FIGURE 2 HPLC chromatogram obtained for benzo[f]quinoline-5,6-dione.

System suitability testing was performed to determine the accuracy and precision of the system from six replicate injections of a solution containing 0.20 mg benzo[f]quinoline-5,6-dione /mL. The relative standard deviation (RSD) of the retention time (min) and peak area were found to be less than 0.25%. The retention factor (also called capacity factor, k) was calculated using the equation $k = (t_r / t_0) - 1$, where t_r is the retention time of the analyte and t_0 is the retention time of an unretained compound; in this study, t_0 was calculated from the first disturbance of the baseline after injection and capacity factor value was obtained 9.03 for benzo[f]quinoline-5,6-dione peak. The separation factor (α) was calculated using the equation, $\alpha = k_2 / k_1$ where k_1 and k_2 are the retention factors for the first and last eluted peaks respectively. The separation factor for benzo[f]quinoline-5,6-dione peak was 2.14 obtained. The plate number (also known as column efficiency, N) was calculated as $N = 5.54 (t_r / w_{0.5})^2$ where $w_{0.5}$ is the peak width at half peak height. In this study, the theoretical plate number was 2826. Resolution is calculated from the equation $R_s = 2(t_2 - t_1) / (t_{w1} + t_{w2})$. Where t_1 and t_2 are retention times of the first and second eluted peaks, respectively, and t_{w1} and t_{w2} are the peak widths. The resolution for benzo[f]quinoline-5,6-dione peak was > 2.0.

For the determination of method robustness within a laboratory, a number of chromatographic parameters were determined which included flow rate, temperature, mobile phase composition, and column from different lots. In all cases good separations were always achieved, indicating that the method remained selective for all components under the tested conditions.

3.2 LC-MS Analysis

LC-MS experiments were performed to identify the chemical structure of impurities using the conditions described in LC-MS conditions section. In my experience the HPLC flow rate had to be decreased to 0.8 mL/min. However, with this low flow rate, while keeping same solvent system and gradient as those for HPLC-PDA method, the retention time will be slightly longer than HPLC-PDA.

The full scan ESI-MS spectra of benzo[f]quinoline-5,6-dione and its impurities were measured in the mass range m/z 2-800 da. The benzo[f]quinoline-5,6-dione displayed a single peak at 9.187 min, which corresponds to the molecular mass at m/z 210.05 (Figure 3) and an impurity peak (12.897 min) at m/z 198.02 (Figure 4) and impurity peak (15.922) at m/z 180.08 (Figure 5), These both impurity peaks at 12.897 and 15.922 min were identified, by both LC and MS data with [M + H]$^+$ at m/z 197.0, as 1h-indeno[2,1-b]pyridine-2,9-dione and [M + H]$^+$ at m/z 179.0, as benzo[f]quinoline (Table 1). Figure 6 shows MS and PDA spectra for impurity peak at (6.844 min) was remaining unidentified.

TABLE 1 Chromatographic results for benzo[f]quinoline-5,6-dione and its impurities

Compound	Chemical name	Rt (min)	Purity (%)	Mass (m/z)
Impurity 1	Unknown	6.844	1.16	Unknown
Benzo[f]quinoline -5,6-dione	-	9.187	89.21	210.05
Impurity 2	1H-indeno[2,1-b]pyridine-2,9-dione	12.897	0.34	198.02
Impurity 3	benzo[f]quinoline	15.922	4.06	180.08

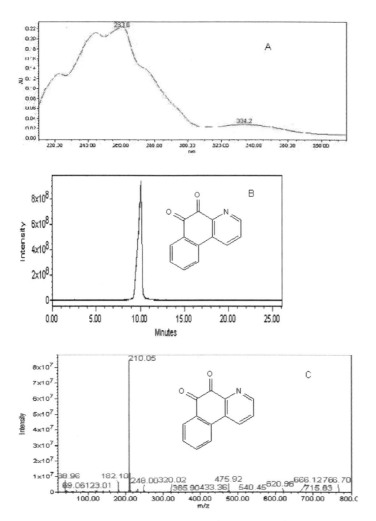

FIGURE 3 (A) PDA UV match spectra of the middle of the peak corresponding to the RT (9.187 min) of the main component of benzo[f]quinoline-5,6-dione and a reference sample; (B) HPLC-ESI-MS chromatogram; (C) mass spectrum (x-axis: relative abundance) obtained for benzo[f]quinoline-5,6-dione.

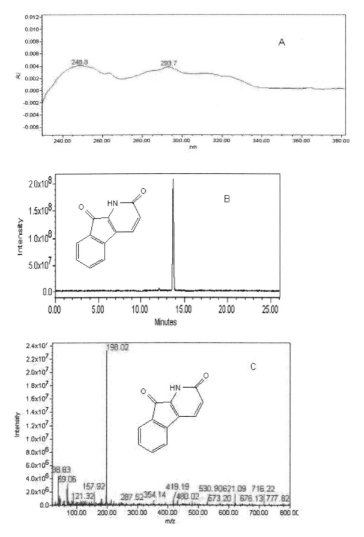

FIGURE 4 (A) PDA UV spectra; (B) HPLC-ESI-MS chromatogram; (C) mass spectrum (x-axis: relative abundance) of the impurity peak corresponding to the RT (12.897 min) obtained for impurity.

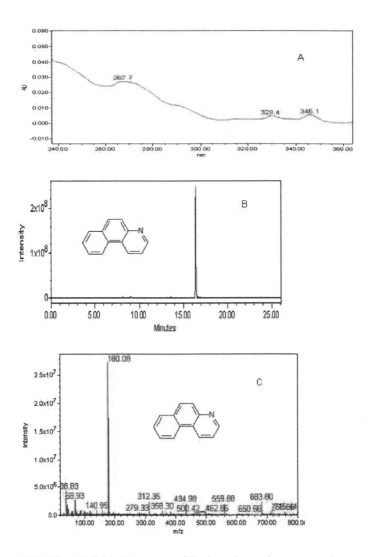

FIGURE 5 (A) PDA UV spectra of the impurity peak corresponding to the RT (15.922 min); (B) HPLC-ESI-MS chromatogram; (C) mass spectrum (x-axis: relative abundance) obtained for impurity.

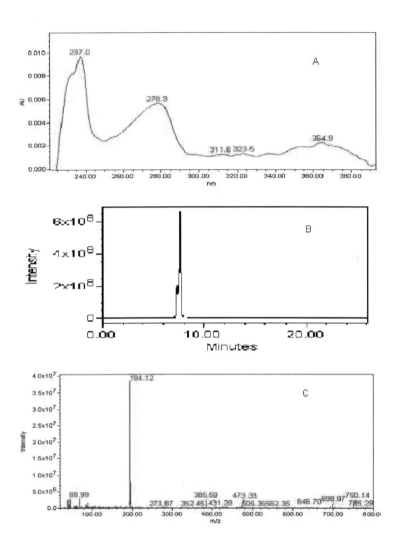

FIGURE 6 (A) PDA UV spectra of the impurity peak corresponding to the RT (6.844 min); (B) HPLC-ESI-MS chromatogram; (C) mass spectrum (x-axis: relative abundance) obtained for impurity.

3.3 Validation of the HPLC Method

The new developed analytical method was critically validated to assess the validity of research means determining whether the method used during the study can be trusted to provide a genuine, account of the intervention being evaluated. As a best practice [13-15], the method was validated in terms of linearity, precision (repeatability and intermediate precision), specificity, accuracy (recovery), limit of detection and limit of quantification.

3.3.1 Linear concentration range

Linearity range of peak area response versus concentration for benzo[f]quinoline-5,6-dione was studied from approximately 0.02 to 1.40 mg/mL. Six solutions were prepared corresponding to 20, 40, 60, 80, 100, and 140% of the nominal analytical concentration (0.20 mg/mL) and each one was injected in triplicate. The correlation coefficient, r^2, was 0.9997, slope 146444 and intercept was 167799 (Table 2).

TABLE 2 Validation results obtained for the HPLC assay of benzo[f]quinoline-5,6-dione

Validation criterion	Concentration range (mg/mL)	Results
Linearity	0.02 to 1.40	$y = 146444x + 167799$
($n = 2$; $k = 6$)		($r^2 = 0.9997$)
Accuracy	0.04	88.33 ± 0.65
(Recovery $\pm$ %RSD; $n = 3$)	0.60	90.33 ± 0.64
	1.40	88.50 ± 0.65
LOD		(s/n = 3.3), 2.0 µg/mL
LOQ ($n = 6$)		(s/n = 10.3), 20.0 µg/mL

3.3.2 Precision (repeatability and intermediate precision)

The precision of the analytical method was evaluated in terms of repeatability (intra-day precision) and intermediate (inter-day) precision. Repeatability was assessed injecting six replicate injections at 100% test concentration (0.20 mg/mL). The %RSD values for peak area and retention time presented in Table 3 was less than 0.57% in each case and illustrated good repeatability precision for the analytical method.

TABLE 3 Repeatability precision study data for benzo[f]quinoline-5,6-dione

Injection	Peak area (μV)	Retention time (min)
1	5308596	9.19
2	5312528	9.19
3	5311452	9.2
4	5302671	9.29
5	5327812	9.19
6	5382762	9.19
Mean	5324303	9.21
RSD (%)	0.56	0.44

Intermediate precision (interday precision) was demonstrated by two analysts using two HPLC systems on different days and evaluating the peak area data across the HPLC systems at three concentration levels (0.04, 0.60 and 1.40 mg/mL) that cover the assay method range 0.02 to 1.40 mg/mL for benzo[f]quinoline-5,6-dione. The mean and %RSD across the HPLC systems and analysts were calculated from the individual peak area mean values at the 0.04, 0.60 and 1.40 mg/mL of the test concentration. The %RSD values for both instruments and analysts were < 0.85% (Table 4) for benzo[f]quinoline-5,6-dione and illustrated the good precision of the analytical method

TABLE 4 Intermediate precision studies data for benzo[*f*]quinoline-5,6-dione

Injection	Analyst 1, day 1, HPLC 1			Analyst 2, day 2, HPLC 2		
	0.04[a]	0.6	1.4	0.04	0.6	1.4
1	5352632[b]	5382372	5376291	5324562	5400236	5316282
2	5318324	5406276	5367286	5362861	5376235	5388117
3	5303271	5367428	5335172	5377242	5312872	5366245
Mean	5324742	5385359	5359583	5354888	5363114	5356881
RSD (%)	0.47	0.36	0.40	0.51	0.84	0.69

[a]Concentration (mg/mL); [b]Peak area (μV s)

3.3.3 Specificity/selectivity

Specificity can also be determined by measurement of peak homogeneity. Because the different techniques available in a PDA are not equally effective for the detection of possible impurities or interference in a chromatographic peak, the use of several techniques is recommended.

In this work the techniques used to validate the peak purity of the benzo[*f*]quinoline-5,6-dione was: (a) Normalization and comparison of spectra from different peak sections, (b) Absorbance of two wavelengths. Both techniques showed that the peak corresponding the benzo[*f*]quinoline-5,6-dione studied present a high level of purity. Also, a chromatogram of the blank was recorded under the same conditions and the signal obtained from this blank showed no interference from any other materials.

Forced degradation studies were also performed to evaluate the specificity of benzo[*f*]quinoline-5,6-dione under four stress conditions (heat, UV light, acid, base). Solutions of benzo[*f*]quinoline-5,6-dione were exposed to 50°C for 1 h, UV light using a Mineralight UVGL-58 light for 24 h, acid (1M hydrochloric acid) for 24 h and base (1M sodium hydroxide) for 4 h. A summary of the stress results is shown in Table 5. It is evident from Fig. 2 that the method has been able to separate the peaks of the degraded products from that of the benzo[f]quinoline-5,6-dione.

TABLE 5 Specificity results of benzo[f]quinoline-5,6-dione under stress conditions

Stress conditions	Sample treatment	Rt (min)	Area (μVs)	Assay (%)
Reference	Fresh solution	9.19	5315628	89.76
Acid	1M HCl for 24 h	9.19	5329526	89.26
Base	1M NaOH for 4 h	9.19	5308364	89.12
Heat	50 °C for 1 h	9.20	5326538	89.01
Light	UV Light for 24 h	9.20	5301672	88.62

3.3.4 Accuracy/recovery studies

In order to test the efficiency of the analytical method, recovery studies at three known added concentration levels (0.04, 0.60 and 1.40 mg/mL) were carried out. Two replicates were prepared at each concentration level and each one was injected in triplicate. Mean recoveries higher than 88% were obtained in all cases with acceptable RSD. The results are shown in Table 2.

3.3.5 Limits of detection and quantification

The limit of detection (LOD) was considered as the minimum analyte concentration yielding a signal-to-noise ratio equal to three. The limit of quantification (LOQ) was adopted at the lowest analyte concentration yielding a signal 10 times greater than the noise. The LOD and LOQ values were for benzo[*f*]quinoline-5,6-dione were found to be 2.0 μg/mL (s/n = 3.3) and 20.0 μg/mL (s/n = 10.3), with RSD 0.88% for six injection replicates, respectively (Table 2).

4. CONCLUSION

A new, sensitive, specific and robust reversed-phase HPLC method for the identity, assay, and purity evaluation of benzo[*f*]quinoline-5,6-dione and its synthetic impurities is described. The research methodology and data for its validity has been extensively validated and showed a good linearity, precision and accuracy. 1*H*-indeno[2,1-*b*]pyridine-2,9-dione and benzo[*f*]quinoline were identified as main impurities by applying the method in LC-MS mode. The developed method could be satisfactorily applied as a routine procedure to identify and quantify benzo[*f*]quinoline-5,6-dione and its synthetic impurities.

REFERENCES

1. Kradolfer, F.; Sachmann, W.; Bassil, G. T. Comparative microbiological studies of amebicides with special reference to the derivatives of phenanthroline quinone. *Ann. Biochem. Exptl. Med.* 1960, *20*, 519-527.
2. Klog, K.; Mlochowski, J.; Szulc, Z. Synthesis of azafluorenones. *J. f. prakt. Chemie* 1977, *319*, 959-967.
3. Zhong, D.; Blume, H. HPLC determination of loratadine and its active metabolite descarboethoxyloratadine in human plasma. *Pharmazie* 1994, *49*, 736-739.

4. Martens, J. Determination of loratadine and pheniramine from human serum by GC-MS. *J. Chromatogr. B* 1995, *673*, 183-188.

5. ICH: Guidelines availability: Impurities in new drug substances: Notice, Fed. Reg. 1996, *61*, 371.

6. Prabhu, S. V.; Ballard, J. M.; Reamer, R. A.; Ellison, D. K. Isolation and identification of impurities in L-696, 229 drug substance. *Talanta*, 1993, *40*, 989-994.

7. Fenn, J.B.; Mann, M.; Meng, C.K.; Wong, S.F.; Whitehouse, C.M. Electrospray ionization for mass spectrometry of large biomolecules. *Science* 1989, *246*, 64-71.

8. Whitehouse, C.M.; Dreyer, R.N.; Yamashita, M.; Fenn, J.B. Electrospray interface for liquid chromatographs and mass spectrometers. *Anal. Chem.* 1985, 57, 675-679.

9. Bruins, A.P.; Covey, T.R.; Henion, J.D. Ion spray interface for combined liquid chromatography/atmospheric pressure ionization mass spectrometry. *Anal. Chem.* 1987, *59*, 2642-2646.

10. Smith, R.D.; Loo, J.A.; Edmonds, et al. Sensitivity considerations for large molecule detection by capillary electrophoresis-electrospray ionization mass spectrometry. *J. Chromatogr.* 1990, *516*, 157-165.

11. Hofstadler, S.A.; Swanek, F.D.; Gale, D.C.; Ewing, A.G.; Smith, R.D. Capillary electrophoresis-electrospray ionization fourier transform ion cyclotron resonance mass spectrometry for direct analysis of cellular proteins. *Anal. Chem.* 1995, *67*, 1477-1480.

12. Lee, E.D.; Mueck, W.; Henion, J.D.; Covey, T.R. Liquid junction coupling for capillary zone electrophoresis/ion spray mass spectrometry. *Biomed. Environ. Mass Spectrom.* 1989, *18*, 844-850.

13. Shabir, G.A.; Lough, J.W.; Arain, S. A.; Bradshaw, T.K. Evaluation and application of best practice in analytical method validation. *J. Liq. Chromatogr. Rel. Technol.* 2007, *30*, 311-333.

14. Shabir, G.A. Validation of HPLC methods for pharmaceutical analysis: Understanding the differences and similarities between validation

requirements of the FDA, USP and ICH. *J. Chromatogr. A* 2003, *987*, 57-66.

15. Shabir, G.A. Step-by-step analytical methods and protocol in the quality system compliance industry. *J. Validation Technol.* 2004, *10*, 314-324.

8 DETERMINATION OF THE SPERMICIDES NONOXYNOL-9 IN A GEL PHARMACEUTICAL FORMULATION USING HPLC

ABSTRACT

The purpose of the research described herein was to develop and validate a reversed-phase high-performance liquid chromatography (RP-HPLC) method for the commercially available spermicides nonoxynol-9 (1.90-2.10%, w/w) in a gel formulation. The chromatographic separation was achieved with methanol-water (83:17, v/v) as mobile phase, a Nucleosil Cyano column, and UV detection at 289 nm. The calibration curve showed good linearity (R^2 = 0.9997) over the concentration range of 0.05-0.35 mg/mL. The recovery ranges were from 99.87-100.04% from a gel formulation. The mean percent relative standard deviation values for intra- and inter-day precision studies were less than 1.0%. The method was also shown to have adequate sensitivity with a detection limit of 0.0065 µg/mL. Acceptable robustness indicates that the assay method remains unaffected by small but deliberate variations, which are described in ICH Q2(R1) guideline.

Keywords: HPLC, Method validation, Nonoxynol-9, Gel formulation, ICH

1. INTRODUCTION

Nonoxynol-9 (nonylphenoxy-polyethylleneoxy-ethanol, $C_{33}H_{60}O_{10}$, Figure 1) is a non-ionic surfactant and one of the most common active spermicidal ingredients in commercially available vaginal contraceptives. Women have used vaginal products in attempts to prevent pregnancy for centuries. Nonoxynol-9 is the only vaginal spermicides currently available in the United States (although other products are marketed in other countries). Products containing nonoxynol-9 have been available over the counter for nearly 50 years.

Nonoxynol-9 is comprised of multiple oligomers due to its synthesis from nonyl phenol and ethylene oxide that vary in ethyleneoxide chain length and consequently in their molecular weights [1]. Nonoxynol-9 is used as multiple oligomers as the active component in various formulations ranging from gels to creams which are administered into the vaginal cavity. The basis of this compound activity is associated with the structural affinity of nonoxynol-9 for membrane lipids, which serves to cause rapid immobilization and cell death by disruption of sperm membrane integrity [2, 3]. Consequently, frequent high dose usage of these contraceptive agents have been reported to have the same effect within the female tract, causing local lesions of the cervicovaginal epithelium and a burning sensation of the genitals of either one or both partners [4]. Furthermore, a recent survey has shown that the existence of such lesions can increase the rate of HIV infection [5-8].

The literature presently describes only one analytical method for the determination of nonoxynol-9 in vaginal lavage fluid using normal-phase LC with bonded phase aminosilica column [9].

The purpose of this study was therefore to develop a simple, sensitive, precise and robust reversed-phase high-performance liquid chromatography (RP-HPLC) method for the determination of nonoxynol-9 in gel formulation.

Analytical method development and validation is an important part of analytical chemistry and plays a major role in the discovery, development, and manufacture of pharmaceuticals. The official test methods that result from these processes are used by quality control laboratories to ensure the identity, purity, potency and performance of drug product 'quality' essential for drug safety and efficacy. RP-HPLC is the analytical method of choice in pharmaceutical analysis because of its specificity and sensitivity. Finally, the developed analytical method was validated to assess the validity of research data means determining whether the method used during the study can be reliable to provide a genuine, account of the intervention being evaluated. As a best practice [10-14], in the subsequent investigation, the new RPLC method was fully validated [15].

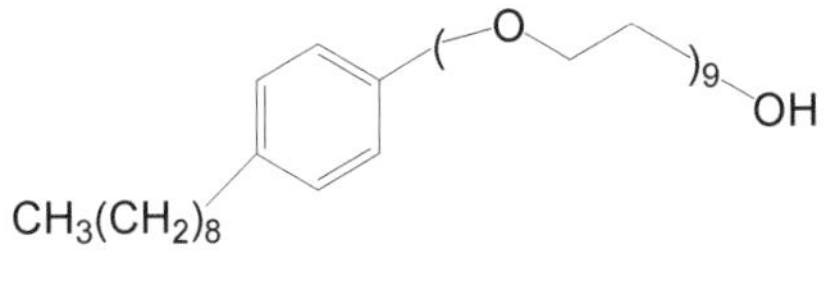

CH$_3$(CH$_2$)$_8$

FIGURE 1 Structure of nonoxynol-9.

2. EXPERIMENTAL

2.1 Chemical and Reagents

Methanol (HPLC-grade) and nonoxynol-9 were obtained from Sigma-Aldrich (Gillingham, UK). Distilled water was de-ionised by using a Milli-Q system (Millipore, Bedford, MA).

2.2 LC instrumentation and Conditions

The Knauer LC system (Berlin, Germany) equipped with a model 1000 LC pump, model 3950 autosampler, model 2600 photodiode-array (PDA) detector and a vacuum degasser was used. The data were acquired via ClarityChrom Workstation data acquisition software. RPLC analysis was performed isocratically at 30°C using a Nucleosil CN (150 × 4.6 mm, 10 µm) column (Jones Chromatography, Hengoed, UK). The mobile phase consisted of a mixture of methanol/water (83:17, v/v) was used. The flow rate was 1.0 mL/min and injection volume was 10 µL. The eluent was monitored with a UV detector at 289 nm.

2.3 Standard Preparation

Nonoxynol-9 (0.2 g) was accurately weighed and added to a 100 mL volumetric flask and dissolved in a HPLC grade methanol (stock solution 1). A 10 mL aliquot of stock solution 1 was diluted to 100 mL in the mobile phase, yielding a final concentration of 0.2 mg/mL.

2.4 Sample Preparation

An accurately weighed amount (1 g) of nonoxynol-9 sample gel was dissolved in 100 mL mobile phase. The sample was filtered through a sample filtration unit (0.45 µm) and injected into the LC system.

3. RESULTS AND DISCUSSION

3.1 Method Development

Initially, two analytical columns were studied in order to reach an acceptable separation, thus specificity and selectivity. Initially, Lichrosorb C_8 (150 × 4.6 mm, 5 µm) column was exploited. The Lichrosorb column gave a poor peak shape with high tailing and longer retention time as 5.35 min (Figure 2) for the analyte peak. Nucleosil CN column (150 × 4.6 mm, 10 µm) produced a peak with superior band shape and column efficiency with much shorter retention time (2.03 min) under the same analytical conditions (Figure 3a).

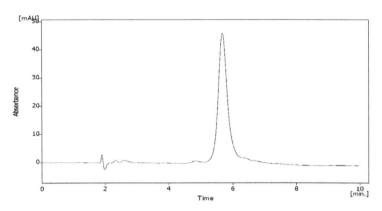

FIGURE 2 RP-HPLC chromatogram of nonoxynol-9 obtained using Lichrosorb C_8 (150 × 4.6 mm, 5 μm) column.

Tests were performed at different temperature between 25°C and 35°C at 5°C steps, to study the influence of this parameter. The results showed that the variation of the temperature between 25°C and 35°C did not significantly affect any of the chromatographic parameters and only decreased the retention time of the analyte so 30°C was selected as the working temperature. The choice of wavelength is essential to accomplish a sensitive chromatographic assay. The optimal wavelength for nonoxynol-9 detection was established using the scan range of 200 to 400 nm. It was established that 289 nm was the optimal wavelength to maximize the signal.

To evaluate the quantitative nature of the analytical method, a series of samples with different amounts of nonoxynol-9 were run to investigate the best assay concentration. Using a Nucleosil CN column, the best concentration was assessed by injecting five reference standards of nonoxynol-9 in the range of 0.05-0.35 mg/mL. The integrated peak areas were plotted versus amount injected. The calibration curve was found to be linear for a concentration range 0.05-0.35 mg/mL with a determination coefficient (r^2) of 0.9997. On the bases of these data, 0.20 mg/mL of the linearity was chosen as the best working concentration for the assay.

System suitability testing was performed by injecting ten replicate injections of a solution containing 0.2 mg nonoxynol-9/mL. The percent relative standard deviation (RSD) of the peak area responses was measured, giving an average of 0.14 ($n = 10$). The tailing factor (T) for each nonoxynol-9 peak was 1.12, the theoretical plate number (N) was 6745, and the retention time (t_R) variation was less than 1% for ten replicate injections. The RPLC method met these requirements within the accepted limits [11, 16].

The stability of nonoxynol-9 in solution containing 0.20 mg/mL of nonoxynol-9 was investigated. The solutions were stable during the investigated 48 h and the percent RSD was in between 0.18 and 0.66% for peak area and height. Standard solutions stored in a capped volumetric flask on a laboratory bench under normal lighting conditions for 48 h, were shown to be stable with no significant change over this period. This is indicated (0.6% changes in area between $T = 0$h and $T = 48$h).

Robustness testing was performed during the method development phase to optimize final RPLC conditions. A RPLC method must prove to be able to remain unaffected by small, but deliberate variations in method parameters, thus showing its own reliability during normal usage. In this study, robustness of the method was evaluated by the analysis of nonoxynol-9 under different experimental conditions such as deliberated changes in the composition of the mobile phase, column temperature and flow rate. The percentage of methanol in the mobile phase was varied ±5%, the column temperature was varied ±5°C and, the flow rate was varied ±0.2 mL/min. Besides, the method was applied using different lots of Nucleosil CN columns. Their effects on the retention time (t_R), tailing factor (T), and theoretical plate numbers (N) were studied as shown in Table 1.

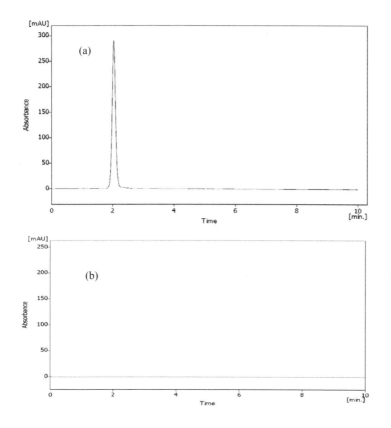

FIGURE 3 RP-HPLC chromatogram obtained from injection of (a) sample; (b) placebo sample using Nucleosil CN (150 × 4.6 mm, 10 μm) column.

TABLE 1 The robustness data of the developed RPLC method

Method parameter	Modification	t_R (min)	T	N
	78	2.09	1.25	6786
Methanol (%)	83	2.03	1.00	6745
	88	2.00	1.05	6648
	0.8	2.08	1.13	6835
Flow rate (mL/min)	1.0	2.00	1.16	6743
	1.2	1.98	1.12	6794
	25	2.01	1.12	6638
Temperature (°C)	30	2.03	1.05	6746
	35	2.00	1.25	6642

3.2 Method Validation

The method was fully validated [15] with respect to linearity, range, accuracy, precision, specificity, limits of detection, limit of quantitation and stability of analytical solutions.

3.2.1 Linearity/range

Calibration graph was established in the range of 0.05-0.35 mg/mL, at the concentrations of 25, 35, 50, 75, 100, 125, 150, and 175%, of the theoretical concentration in the test preparation, n = 3). Standard solutions of nonoxynol-9 were prepared by serial dilution in mobile phase to yield concentrations of 0.05, 0.07, 0.10, 0.15, 0.20, 0.25, 0.30, and 0.35 mg/mL. For calibration, weighed linear regression analysis was applied. The calibration graph established was linear (r^2 = 0.9997) within the tested range of 0.05-0.35 mg/mL (Table 2).

TABLE 2 Linearity assessment of the HPLC method for the assay of nonoxynol-9

Concentration (mg/mL)	Concentration as percent of 0.20 mg/mL of nonoxynol-9	Peak area (mAU.s)	RSD (%)	n
0.050	25	897	0.26	3
0.070	35	1302	0.47	3
0.100	50	1838	0.49	3
0.150	75	2769	0.65	3
0.200	100	3702	0.28	3
0.250	125	4655	0.32	3
0.300	150	5638	0.25	3
0.350	175	6669	0.62	3

Determination coefficient (r^2): 0.9997
Equation for regression line: $y = 18973x - 46.663$

3.2.2 Stability of analytical solutions

Samples and standard solutions were chromatographed immediately after preparation and then re-assayed after storage at room temperature for 48 h. The results given in Table 3 show that there was no significant change (<1% response factor) in nonoxynol-9 concentration over this period.

3.2.3 Precision

The precision of the method was determined by repeatability (intra-day) and intermediate precision (inter-day variation). The repeatability (intra-day precision) of the method was evaluated by assaying six replicate injections of the nonoxynol-9 at 100% of test concentration (0.2 mg/mL). The %RSD of the retention times (min) and areas of nonoxynol-9 peak were found to be less than 0.14% (Table 3).

Intermediate precision (inter-day variation) was demonstrated by two analysts using two LC systems and evaluating the relative peak area percent data across the two LC systems at three concentration levels (75%, 100%, and 125%) that cover the assay method range (0.05-0.35 mg/mL). The mean and %RSD across the systems and analysts were calculated from the individual relative percent peak area mean values at the 75%, 100%, and 125% of the test concentration. The %RSD values for both instruments and analysts were less than 0.26% (Table 3) and illustrated the good precision for the analytical method.

TABLE 3 Validation results of nonoxynol-9

Validation step	Parameters	Acceptance criteria	Results
Standard stability	% change in response factor	< 2%	0.09
Sample stability	% change in response factor	< 2%	0.11
Repeatability	t_R (min) %RSD	< 2%	0.09
($n = 6$)	Peak area %RSD		0.13
Intermediate	Instrument %RSD	< 2%	0.22
precision	Analyst %RSD	< 2%	0.25
($n = 6$)			
LOD	µg/mL	-	0.0065
LOQ	µg/mL	-	9.4
System suitability	Peak area %RSD	< 2%	0.12
	Retention times (min) %RSD	< 2%	0.09
	Tailing factor	≤ 2	1.11
	Theoretical plates	> 2000	6758

3.2.4 Accuracy/recovery studies

The accuracy of the method was evaluated by means of recovery assay, adding known amounts of nonoxynol-9 reference standard to a known amount of gel formulation in order to obtain three different levels (75%, 100%, and 125%) of addition. The samples were analysed and the mean recovery was calculated. The data presented in Table 4 shows the recovery of nonoxynol-9 in spiked samples met the evaluation criteria for accuracy (100 +/-2.0% over the range of 80 to 120% of target concentration).

TABLE 4 Recovery studies of nonoxynol-9 from samples with known concentration

Sample #	Percent of nominal	Amount of nonoxynol-9 (µg/mL)		Recovery (%) ($n = 3$)	Bias %	RSD (%) ($n = 3$)
		Added	Obtained			
1	75	150	149.8	99.87	-11	0.18
2	100	200	199.8	99.90	-12	0.22
3	125	250	250.2	100.08	-10	0.29
Mean				99.95		

3.2.5 Specificity

The RPLC-PDA/UV isoplot chromatogram (Figure 4) demonstrates a good separation of the nonoxynol-9. The isoplot chromatogram data consist of PDA UV-vis absorption spectra from 200 to 400 nm for each point along the chromatogram. Injections of the extracted placebo were also performed to demonstrate the absence of interference with the elution of the nonoxynol-9. These results demonstrate (Figure 3b) that there was no interference from the other materials in the gel formulation and, therefore confirm the specificity of the RPLC method.

The forced degradation studies were performed to evaluate the specificity of nonoxynol-9 under four stress conditions (heat, UV light, acid, base). Solutions of nonoxynol-9 were exposed to 60°C for 1 h, UV light using a UVL-56 lamp for 24 h, acid (1M hydrochloric acid) for 24 h and base (1M sodium hydroxide) for 4 h. A summary of the stress results is shown in Table 5. No significant degradation was observed under any stress conditions studied and therefore confirm the specificity of the RPLC method.

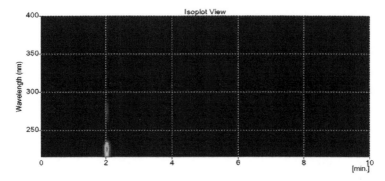

FIGURE 4 RPLC-PDA/UV isoplot chromatogram of nonoxynol-9.

Table 5 Assay (%) of nonoxynol-9 under stress conditions

Stress conditions	Sample treatment	Concentration (mg/mL)	t_R (min)	Assay (%)	Area (mAU s)
Reference	Fresh solution	0.2	2.03	99.6	2414.18
Acid	1M HCl for 24 h	0.2	2.05	100.0	2417.23
Base	1M NaOH for 4 h	0.2	2.05	98.9	2346.14
Heat	60 °C for 1 h	0.2	1.97	98.8	2376.22
Light	UV Light for 24 h	0.2	2.00	99.9	2402.38

3.2.6 Limits of detection and quantitation

The limit of detection (LOD) and limit of quantitation (LOQ) of nonoxynol-9 was determined based on standard deviation (σ) of response and slope (s). Nonoxynol-9 solutions were prepared in the range 0.004-150 µg/mL and injected in triplicate. Average peak area of analyte was plotted against concentration. LOD and LOQ were calculated by using the following equations:

LOD = (3.3 σ)/s

LOQ = (10 σ)/s

The LOD was determined to be 0.0065 µg/mL and LOQ was found to be 9.4 µg/mL for nonoxynol-9 with %RSD less than 0.24% for six replicate injections (Table 3).

3.2.7 System suitability test

A system suitability test was performed by injecting six replicate injections of a solution containing 0.2 mg nonoxynol-9/mL. The RSD of the peak area responses was measured, giving an average of 0.12% ($n = 6$). The tailing factor (T) for each nonoxynol-9 peak was 1.11, the theoretical plate number (N) was 6758, and the retention time (t_R) variation was less than 0.09% for six replicate injections as shown in Table 3.

4. CONCLUSION

This is the first RPLC method developed and validated for the determination of nonoxynol-9 in gel formulation using UV detection. A simple isocratic mobile phase was used without adding buffer. Sample preparation and analytical procedure run times are short (<3 min). The proposed method for its validity was validated and results showed excellent linearity (R^2 = 0.9997), good precision (RSD < 1.0%) and excellent recovery (>99.8%). This proposed validated robust method can be reliably used in routine analysis in quality control for release of raw materials, bulk gel samples and finished products congaing nonoxynol-9 compound.

ACKNOWLEDGEMENT

I thank Dr Tony Bradshaw (Oxford Brookes University) for his valued advice on this work

REFERENCES

1. Walter, B.A., Hawi, A.A., Zavou, P.M., Digenis, G.A., Solubilization and invitro spermicidal assessment of nonoxynol-9 and selected fractions using rabbit spermatozoa. *Pharm. Res.*, 3, 403-408, 1991.
2. Halpern, V., Rountree, W., Raymond, E.G., Law, M. The effects of spermicides containing nonoxynol-9 on cervical cytology. *Contraception*, 77(3), 191-194, 2008.
3. Wilborn, W. H., Hahn, D. W., McGuire, J. J. Scanning electron microscopy of human spermatozoa after incubation with the spermicide nonoxynol-9. *Fertil. Steril.*, 39, 717–719, 1983.

4. Roddy, R.E., Cordero, M., Cordero, C., Formey, J.A.A. Current opinion in infectious diseases. *Int. J. STD AIDS*, 4, 163-165, 1993.

5. Wilkinson, D. Nonoxynol-9 fails to prevents STDs, but microbicide research continues. *Lancet*, 360(9338), 962-963, 2002.

6. Roddy, R.E., Zekeng, L., Ryan, K.A., Tamoufe, U., Tweedy, K.G. Effect of nonoxynol-9 gel on urogenital gonorrhea and chlamydial infection: a randomized controlled trial, *JAMA.*, 287(9), 1117-22, 2002.

7. Van, D. L., Ramjee, G., Alary, M., et.al. Effectiveness of COL-1492, a nonoxynol-9 vaginal gel, on HIV-1 transmission in female sex workers: a randomised controlled   trial. *Lancet*, 360(9338), 971-977, 2002.

8. Roddy, R.E., Zekeng, L., Ryan, K.A., Tamoufe, U., Weir, S.S., Wong, E.L. A controlled trial of nonoxynol 9 film to reduce male-to-female transmission of sexually transmitted diseases. *N. Engl. J. Med.*, 339(8), 504-510, 1998.

9. Jason, L. M., James, H.N., Patricia, B.C., Fayez, M.H. Determination of the spermicide nonoxynol-9 in vaginal lavage by HPLC. *J. Chromatogr. B.*, 677(1), 204-208, 1996.

10. Shabir, G.A., Lough, W.J., Shafique, A.A., Bradshaw, T.K. Evaluation and application of best practice in analytical method validation. *J. Liq. Chromatogr. Relat. Technol.*, 30 (3), 311-333, 2007.

11. Shabir, G.A. Validation of HPLC methods for pharmaceutical analysis: Understanding the differences and similarities between validation requirements of the U.S. Food and Drug Administration, the U.S. Pharmacopoeia and the International Conference on Harmonization. *J. Chromatogr. A.*, 987(1-2), 57-66, 2003.

12. Shabir, G.A. Step-by-step analytical methods and protocol in the quality system compliance industry. *J. Validation Technol.*, 10(4), 314-324, 2004.

13. Shabir, G. A. A practical approach to validation of HPLC methods under current good manufacturing practices. *J. Validation Technol.*, 10(3), 210-218, 2004.

14. Shabir, G. A. HPLC method development and validation for pharmaceutical analysis. *Pharma. Technol. Eur.*, 16(3), 37-49, 2004

15. International Conference on Harmonization (ICH), Validation of analytical procedures: Text and Methodology, Q2 (R1), Geneva, Switzerland, Nov. 2005.

16. U.S. FDA. Reviewer Guidance: Validation of Chromatographic Methods, Center for Drug Evaluation and Research (CDER), 1994.

9 DETERMINATION OF 2-(DIETHYLAMINO)-N-(2,6 DIMETHYLPHENYL) ACETAMIDE IN A GEL PHARMACEUTICAL FORMULATION BY HPLC

ABSTRACT

An isocratic reversed-phase high-performance liquid chromatographic (HPLC) method was developed and validated for the determination of 2-(diethylamino)-N-(2,6-dimethylphenyl) acetamide (0.3% w/w) in a gel formulation. The chromatographic separation was achieved with potassium phosphate buffer (pH 7.0)-acetonitrile (47:53, v/v) as mobile phase, a C_{18} column and UV detection at 254 nm. The calibration curve was linear (r^2 = 1.000) from 20-140% range of the analytical concentration of 1.4 μg/mL. The mean %RSD values for intra- and inter-day precision studies were <1%. The recovery ranged between 99.95-100.23% from a gel formulation. The method was specific and successfully routinely used in quality control for the analysis of bulk gel samples and final product release.

Keywords: HPLC, Method validation, 2-(Diethylamino)-N-(2,6-dimethylphenyl) acetamide, Gel formulation

1. INTRODUCTION

2-(Diethylamino)-N-(2,6-dimethylphenyl) acetamide (Figure 1) is widely used as a local anaesthetic that can be administered in a gel matrix [1]. It has also achieved prominence as antiarrhythmic agent and is now in common use particularly as emergency treatment for ventricular arrhythmias that are encountered after cardiac surgery or acute myocardial infection. Some methods for the determination of 2-(diethylamino)-N-(2,6-dimethylphenyl) acetamide in gel formulations have been reported, such as spectrophotometry [2-4], gas liquid chromatography [5] and HPLC [6-9], but method validation has not been reported.

Analytical methods validation is an important regulatory requirement in pharmaceutical analysis. In recent years, the International Conference on Harmonization (ICH) has introduced guidelines for analytical methods validation [10] in Japan, Europe and United States. The most widely applied analytical performance characteristics are accuracy, precision (repeatability and intermediate precision), specificity, limit of detection, limit of quantitation, linearity, range, and stability of analytical solutions. The purpose of this study was to develop and validate a rapid, accurate, sensitive and simple HPLC method for the quantitation of 2-(diethylamino)-N-(2,6-dimethylphenyl) acetamide in a gel pharmaceutical formulation for bulk and final product release.

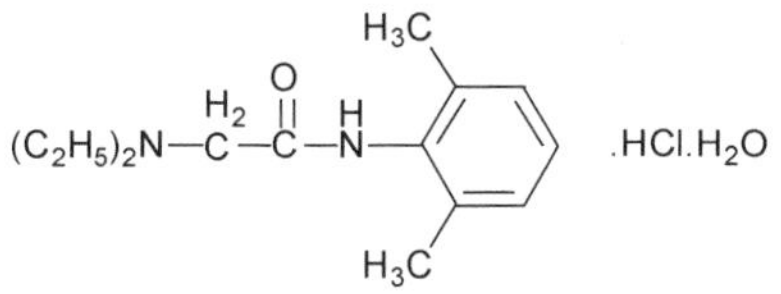

FIGURE 1 Chemical structure of 2-(diethylamino)-N-(2,6-dimethylphenyl) acetamide.

2. EXPERIMENTAL

2.1 Chemicals and Reagents

All chemicals and reagents were of the highest purity. HPLC-grade acetonitrile, 2-(diethylamino)-N-(2,6-dimethylphenyl) acetamide, and potassium phosphate (KH_2PO_4) were obtained from Merck (Darmstadt, Germany). De-ionised distilled water was used throughout the experiment.

2.2 HPLC Instrumentation and Conditions

A Perkin Elmer HPLC system equipped with a model series 200 UV Visible detector, series 200 LC pump, series 200 autosampler and series 200 peltier LC column oven were used to chromatograph the solutions. The data were acquired via PE TotalChrom Workstation data acquisition software, (Version 6.2.0) using PE Nelson series 600 LINK interfaces. The second instrument used in this study was also a Perkin Elmer HPLC system. The mobile phase consisted of a mixture of a potassium phosphate buffer (pH 7.0) and acetonitrile (47:53, v/v). The flow rate was set to 2.0 mL/min and the oven temperature to 25°C. The injection volume was 20 μL and the detection wavelength was set at 254 nm. The chromatographic analysis was carried out on a 3.9 × 300 mm ID, 5 μm C$_{18}$ μ-Bondapak column obtained from Waters (Milford, MA, USA).

2.3 Standard Preparation

An accurately weighed amount (2.8 mg) of 2-(diethylamino)-N-(2,6-dimethylphenyl) acetamide reference standard was placed in a 100 mL volumetric flask and dissolved in de-ionized water (stock). A 5.0 mL aliquot of stock solution was diluted to 100 mL in buffer pH 7.0, yielding a final concentration of 1.4 μg/mL.

2.4 Sample Preparation

An accurately weighed amount (400 mg) of sample gel was dissolved in 100 mL of buffer pH 7.0 to provide a concentration of 4000 μg/mL.

3. RESULTS AND DISCUSSION

3.1 Method Development

The chromatographic analysis of 2-(diethylamino)-N-(2,6-dimethylphenyl) acetamide (pKa 7.86) was carried out in the isocratic mode using a mixture of 53% acetonitrile in buffer pH 7.0 (53:47, v/v) as mobile phase. The column was equilibrated with the mobile phase flowing at 2.0 mL/min for 1 hour prior to injection. The column temperature was ambient. 20 µL of standard and sample solutions were injected automatically into the column. Subsequently, the liquid chromatographic behaviours of both drugs were monitored with a UV detector at 254 nm.

Additionally, preliminary precision, linearity and robustness studies performed during the development of the method showed that the 20 µL injection volume was reproducible and the peak response was significant at the analytical concentration chosen. Diluting the standard and sample in buffer pH 7.0 gave solutions that could be injected directly (without further dilution, filtration or centrifugation). Chromatograms of the 2-(diethylamino)-N-(2,6-dimethylphenyl) acetamide gave good peak shape (Figure 2) and co-elution of excipients was not observed (Figure 3) at the same retention time as 2-(diethylamino)-N-(2,6-dimethylphenyl) acetamide. The retention time for 2-(diethylamino)-N-(2,6-dimethylphenyl) acetamide was 2.12 minutes. System suitability testing was performed by injecting six replicate injections of a solution containing 1.4µg 2-(diethylamino)-N-(2,6-dimethylphenyl) acetamide/mL. The percent relative standard deviation (%RSD) of the peak area responses was measured, giving an average of 0.10 ($N = 6$). The tailing factor (T) for each 2-(diethylamino)-N-(2,6-dimethylphenyl) acetamide peak was 1.05; the theoretical plate number (N) was 8735 and retention time (RT) variation %RSD <1% for six injections.

For the determination of method robustness within a laboratory, a number of chromatographic parameters were determined which included flow rate, temperature, mobile phase composition, and column from different lots. In all cases good separations of 2-(diethylamino)-N-(2,6-dimethylphenyl) acetamide were always achieved, indicating that the method remained selective for 2-(diethylamino)-N-(2,6-dimethylphenyl) acetamide component under the tested conditions.

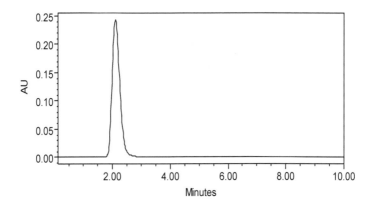

FIGURE 2 HPLC chromatogram of the 2-(diethylamino)-N-(2,6-dimethylphenyl) acetamide.

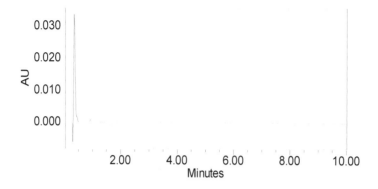

FIGURE 3 HPLC chromatogram of placebo without adding active analyte.

3.2 Validation of the Method

3.2.1 Stability of analytical solutions

Sample and standard solutions were chromatographed immediately after preparation and then re-assayed after storage at room temperature for 48 hours. The results given in Table 1 showed that there was no significant change (< 2% response factor) in 2-(diethylamino)-N-(2,6-dimethylphenyl) acetamide concentration over this period.

3.2.2 Linearity

Linearity was studied using seven different amounts of 2-(diethylamino)-N-(2,6-dimethylphenyl) acetamide in the range of 20-140% around the theoretical values (1.40 µg/mL) and the following equation was found by plotting peak area (y) versus concentration (x) expressed in µg/mL: $Y = 1.30664e+003x$ ($r^2 = 1.000$). The determination coefficient (r^2) obtained (Table 1) for the regression line demonstrates the excellent relationship between peak area and the concentration of 2-(diethylamino)-N-(2,6-dimethylphenyl) acetamide.

3.2.3 Precision (repeatability and intermediate precision)

The precision of the method was investigated with respect to repeatability and intermediate precision. Repeatability (intra-day precision) of the method was evaluated by assaying six replicate injections of the 2-(diethylamino)-N-(2,6-dimethylphenyl) acetamide at 100% of test concentration (1.4 µg/mL). The %RSD of the retention time (min) and relative percent peak area were found to be less than 0.19% (Table 1).

Intermediate precision (inter-day precision) was demonstrated by two analysts, using two HPLC systems and evaluating the relative peak area percent data across the two HPLC systems at three concentration levels (60%, 100%, 120%) that cover the assay method range (0.0002-0.014 g/mL). The mean and %RSD across the systems and analysts were calculated from the individual relative percent peak area mean values at the 60, 100, and 120% of the test concentration. The %RSD values for both instruments and analysts ware <0.13 (Table 1) and illustrated the good precision of the analytical method.

TABLE 1 Method validation results

Validation steps	Parameter	Acceptance Criteria	Results
Standard stability	% Change in response factors	< 2	0.11
Sample stability	% Change in response factors	< 2	0.16
Repeatability (n = 6)	Retention time (min) %RSD	≤ 2	0.01
	Peak area %RSD	≤ 2	0.18
Intermediate precision (n = 3)	Instruments %RSD	≤ 2	0.12
	Analysts %RSD	≤ 2	0.07
Linearity (n = 7)	Correlation coefficient (r^2)	> 0.998	r^2 = 1.000
	Intercept	-8.943 – 6.174	0.142
	Slope	0.93 – 1.081	0.00076
LOD	Signal-to-noise ratio	S/N = 3:1	(S/N = 3.2), 100 ηg/mL
LOQ	Signal-to-noise ratio	S/N = 10:1	(S/N = 10.2), 250 ηg/mL
System suitability (n = 6)	Peak area %RSD	< 2	0.10

3.2.4 Specificity/selectivity

Injections of the extracted placebo were performed to demonstrate the absence of interference with the elution of the 2-(diethylamino)-N-(2,6-dimethylphenyl) acetamide. These results demonstrate (Figure 3) that there was no interference from the other materials in the gel formulation, and therefore confirm the specificity of the method.

3.2.5 Accuracy/recovery studies

The accuracy of the method was evaluated by adding known quantities of 2-(diethylamino)-N-(2,6-dimethylphenyl) acetamide in the gel formulation samples to give a range of 2-(diethylamino)-N-(2,6-dimethylphenyl) acetamide concentration of 75-150% (n = 3) of that in a test preparation. These solutions were analysed and the amount of analyte recovered calculated. The recovery data expressed as an average percent of triplicate injections are presented in Table 2 and shows good recovery of 2-(diethylamino)-N-(2,6-dimethylphenyl) acetamide.

3.2.6 Limits of detection and quantitation

The limit of detection (LOD) and limit of quantitation (LOQ) tests for the procedure were performed on samples containing very low concentrations of analyte. LOD is defined as the lowest amount of analyte that can be detected above baseline noise, typically, three times the noise level. LOQ is defined as the lowest amount of analyte which can be reproducibly quantitated above the baseline noise, that gives S/N = 10. The LOD was (S/N ratio 3.2) 100 ng/mL and the LOQ was (S/N ratio 10.2) 250 ng/ mL and %RSD 0.36% (n = 3).

TABLE 2 Recovery studies of 2-(diethylamino)-N-(2,6-dimethylphenyl) acetamide from samples with known concentration

Sample	Percent of nominal	Amount of analyte (mg)		Recovery (%) (n = 3)	RSD (%) (n = 3)
		Added	Recovered		
1	75	1.70	1.704	100.23	0.13
2	100	3.20	3.204	100.14	0.20
3	150	5.10	5.097	99.95	0.15
Mean				100.10	

4. CONCLUSION

A new HPLC method for the assay of 2-(diethylamino)-N-(2,6-dimethylphenyl) acetamide was developed and fully validated. The results showed that the method is very selective no significant interfering peak was detected; accurate with the percentage recoveries of 99.95-100.23 and reproducible with the %RSD of <1%. The method is sensitive as little as 100 ηg/mL can be detected with the quantitation limit of 250 ηg/mL. The method was used in quality control for analysis of 2-(diethylamino)-N-(2,6-dimethylphenyl) acetamide in bulk, raw materials and final gel products pharmaceutical formulations.

REFERENCES

1. Stegman, M.B., Stoukides, C.A., Resolution of tumor pain with EMLA cream, Pharmacotherapy, 16: 694 (1996).
2. A.G. Gilman, L.S. Goodman, T.W. Rall, F. Murad, Goodman and Gilman's, The pharmacological Basis of Therapeutics, 7th ed. Macmillan, New York, 1980, pp. 310, 767.
3. Fayez, M., El-Tarras, M., Zeinab, S. On the analysis of some local anesthetics by the acid-dye technique. Chemical, Biomedical and Environmental Instrumentation, 11(5-6), 411-423 (1981).
4. G. Ezzat, M. Soad. Lidocaine determination in pharmaceutical preparations. *Egypt. J. Pharm. Sci.* 18: 355-366 (1977).
5. Y. Chen, J.M. Potter, P.J. Ravenscroft, *J. Chromatogr.* 574: 361-364 (1992).
6. Wilson TD, Forde MD. Stability of milrinone and epinephrine, altropine sulphate, lidocaine hydrochloride, or morphine sulphate injection. *Am. J. Hosp. Pharm.* 47: 2504-2507 (1990).

7. Klein J; Fernandes D; Gazarian M; Kent G; Koren G. Simultaneous determination of lidocaine, prilocaine and the prilocaine metabolite o-toluidine in plasma by high-performance liquid chromatography. *J. chromatogr. B,* Biomedical applications, 655(1), 83-88 (1994).

8. Gupta VD, Stewart KR. Chemical stabilities of lignocaine hydrochloride and phenylephrine hydrochloride in aqueous solution. *J Clin. Hosp. Pharm.* 11: 449-452 (1986).

9. Waraszkiewicz, Sigmund M.; Milano, E. A.; DiRubio, R. Stability-indicating high-performance liquid chromatographic analysis of lidocaine hydrochloride and lidocaine hydrochloride with epinephrine injectable solutions. *J. Pharm. Sci.,* 70(11), 1215-1218 (1981).

10. International Conference on Harmonisation (ICH), Q2(R1): Validation of Analytical Procedures: Text and Methodology, Nov. 2005.

10 DEVELOPMENT AND VALIDATION OF A HPLC METHOD FOR THE DETERMINATION OF METHYL SALICYLATE IN A MEDICATED CREAM FORMULATION

ABSTRACT

A new reversed-phase high-performance liquid chromatographic (RP-HPLC) method for the determination of methyl salicylate in a medicated cream formulation was developed and validated. The separation was achieved using an isocratic mobile phase, on a Lichrosorb C_8 column. The eluent was monitored by photodiode array detection at 304 nm. The calibration curve showed excellent linearity (R^2 = 0.9999) over the concentration range of 25-175 µg/mL. The recovery of methyl salicylate was in the range from 99.78-100.0%. The percent relative standard deviation values for intra- and inter-day precision studies were <2.0%. The method is very simple, sensitive and robust with short runtime (<3.0 min) to enable the processing of numerous quality control samples.

Keywords: Methyl salicylate, Reversed-phase LC, Method validation, Medicated cream formulation

1. INTRODUCTION

Methyl salicylate (2-hydroxybenzoic acid methyl ester, methyl 2-hydroxybenzoate, oil of wintergreen) is a naturally occurring compound which can be found in wintergreen oil as well as in sweet Birch. Methyl salicylate (Figure 1) is a non-steroid analgesic and anti-inflammatory drug used in many medicinal formulations for over-the-counter products including muscle ache creams and ointments. Furthermore, the United States Departments of Homeland Security (DHS) and Defense are testing next generation personal protection systems to protect U.S. military personnel from chemical threats. Some of these programs focus on man-in-simulant testing, where methyl salicylate is used to test the effectiveness of chemical suits [1, 2].

Methyl salicylate has two functional groups, the alcohol OH group and the ester group ($COOCH_3$). Esters are formed by the combination of an alcohol, e.g. R-OH and a carboxylic acid, (R-COOH). In pure form, methyl salicylate is toxic, especially when taken internally. A seventeen year-old cross-country runner at the Notre Dame Academy on Staten Island, died April 3, 2007, after her body absorbed high levels of methyl salicylate through excessive use of topical muscle-pain relief products [3]. For these reasons it is essential to analyse and quantify the methyl salicylate drug, which can be considered as a major part of the quality control final product release. The literature presently describes only one analytical method for the determination of methyl salicylate in combination with camphor and menthol in ointment using gas chromatography [4].

The objective of this study was therefore to develop a simple, sensitive, robust and precise reversed-phase high-performance liquid chromatographic (RP-HPLC) method for the determination of methyl salicylate in medicated cream formulation. Frequently, HPLC is the analytical method of choice in pharmaceutical analysis because of its specificity and sensitivity. As a best practice [5-9] the new RP-HPLC method was validated accordance to International Conference on Harmonization (ICH) [10] and U.S. Food and Drug Administration (FDA) [11] guidelines.

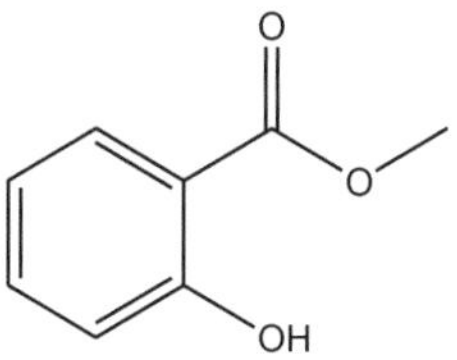

FIGURE 1 Chemical structure of methyl salicylate.

2. EXPERIMENTAL

2.1 Chemicals and Reagents

Methanol (HPLC grade), acetic acid (analytical grade) and methyl salicylate (pure ≥ 99%) were obtained from Sigma-Aldrich (Gillingham, UK). Distilled water was de-ionised by using a Milli-Q system (Millipore, Bedford, MA).

2.2 LC System and Conditions

The Knauer (Berlin, Germany) HPLC system equipped with a model 1000 LC pump, model 3950 autosampler, model 2600 photodiode-array (PDA) detector and a vacuum degasser was used. The data were acquired via Knauer ClarityChrom Workstation data acquisition software. The mobile phase consisted of a mixture of methanol-water (65:35, v/v) containing 1.0% acetic acid. The flow rate was set to 1.0 mL/min. The injection volume was 20 µL and the detection wavelength was set at 304 nm. Reversed-phase LC analysis was performed isocratically at 30 ±0.5°C using a Lichrosorb C_8 (150 mm × 4.6 mm, 5 µm) column (Jones Chromatography, Hengoed, UK).

2.3 Standard Preparation

Methyl salicylate (0.1 g) was accurately weighed and added to a 100 mL volumetric flask before being dissolved in methanol. A 10 mL aliquot of stock solution was diluted to 100 mL in the mobile phase, yielding a final concentration of 100 µg/mL. Standard solutions for the evaluation of methyl salicylate linearity were prepared over a concentration range of 25-175 µg/mL, to 25, 50, 75, 100, 125, 150 and 175%, in the mobile phase.

2.4 Sample Preparation

Approximately (1.0 g) sample from final product commercially available was weighed into a 100 mL volumetric flask and twenty milliliter of methanol was added and flask was heated on a water bath up to boiling point. Then sample was cooled to room temperature and diluted in 100 mL mobile phase. The sample was filtered through 0.45 µ membrane filter and injected into RP-HPLC.

3. RESULTS AND DISCUSSION

3.1 Method Development

The chromatographic separation of methyl salicylate was carried out in isocratic mode using a mixture of methanol-water (65:35, v/v) containing 1.0% acetic acid as mobile phase. The column was equilibrated with the mobile phase flowing at 1.0 mL/min for about 20 min prior to injection. The column temperature was held at 30 ±0.5°C. Twenty microliter of standard solution was injected automatically into the column. Subsequently, the liquid chromatographic behavior of methyl salicylate was monitored with a photodiode-array UV detector at 200-400 nm and signal optimized at 304 nm (Figure 2). Additionally, preliminary system suitability, precision, linearity, robustness and stability of solutions studied performed during the development of the method showed that the 20 µL injection volume was reproducible and the peak response was significant at the analytical concentration chosen. Chromatogram of the resulting solution gave excellent separation (Figure 3).

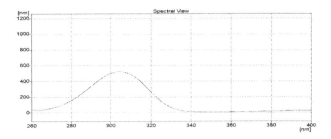

FIGURE 2 PDA UV spectra of the middle of the peak corresponding to the retention time of the main component of methyl salicylate.

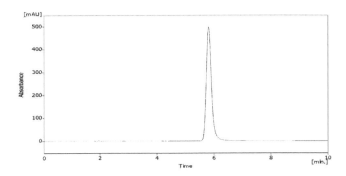

FIGURE 3 HPLC chromatogram of methyl salicylate.

3.1.1 Effect of temperature

Tests were made by varying the temperature between 25°C and 50°C at 5°C steps, to study the influence of this parameter. The results showed that the variation of the temperature between 25°C and 35°C did not significantly affect any of the chromatographic parameters and only decreased the retention time of the analyte (Figure 4) so 30°C was selected as the working temperature.

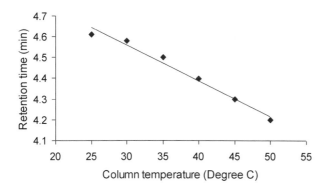

FIGURE 4 Effect of analytical column temperature on the retention times.

3.1.2 System suitability testing

System suitability test was developed for the routine application of the assay method. Prior to each analysis, the chromatographic system must satisfy suitability test requirements (resolution and repeatability). System suitability test was performed to determine the accuracy and precision of the system from six replicate injections of solutions containing 100 µg methyl salicylate per mL. All peaks were well resolved and the precision of injections for methyl salicylate peaks were acceptable. The percent relative standard deviation (RSD) of the peaks area responses were measured, giving an average 0.16% ($n = 6$). The tailing factor (T), capacity factor (K), and theoretical plate number (N) were also calculated. The results of system suitability in comparison with the required limits are shown in Table 1. The proposed method met these requirements within the accepted limits [11, 12].

TABLE 1 System suitability results of the proposed analytical method for methyl salicylate

Parameter	Recommended limits	Results
Retention time (min)	-	5.82
Injection repeatability ($n = 6$)	RSD $\leq$ 1 % for $n \geq 5$	0.16
Capacity factor (K)	K >2	4.02
Tailing factor (T)	T $\leq$2	1.12
Plate number (N)	N >2000	4686

3.1.3 Robustness

To determine the robustness of method, the final experimental conditions were purposely altered and the results were examined. The flow rate was varied by 1 ±0.2 mL/min. the percentage of organic modifier was varied by 65 ± 5% and column temperature was varied by 30 ±5°C. Their effects on the retention time (TR), tailing factor (T), theoretical plate numbers (N) and repeatability of peak areas ($n = 3$) were studied. It can be seen that every employed condition, the chromatographic parameters are in accordance with established value [11]. A change of mobile phase composition, flow rate and temperature had no impact on chromatographic performance. The tailing factor for methyl salicylate was found to be less than 1.3 and analyte was well separated under all the changes carried out (Table 2). Considering the result of modifications in the system suitability parameters and the specificity of the method, it would be concluded that the method conditions are robust.

TABLE 2 Robustness data of the developed RPLC method for methyl salicylate

Parameter	Modification	TR (min)	T	N	RSD (%)*
Mobile phase (Methanol, ± 5%)	60%	6.49	1.22	4614	0.14
	65%	5.82	1.11	4686	0.12
	70%	5.42	1.02	4566	0.17
Flow rate (± 0.2, mL/min)	0.8	5.67	1.11	4688	0.10
	1.0	5.82	1.10	4686	0.09
	1.2	5.53	1.04	4682	0.06
Temperature (± 5°C)	25	6.14	1.14	4689	0.15
	30	5.82	1.12	4686	0.10
	35	5.47	0.98	4654	0.13

*Peak area ($n = 3$)

3.1.4 Stability of analytical solutions

The stability of methyl salicylate in solution containing 0.1 mg/mL of methyl salicylate was investigated. The solutions were stable during the investigated 48 h and the RSD was in between 0.08 and 0.32% for retention time (min), peak area and height. Standard solutions stored in a capped volumetric flask on a laboratory bench under normal lighting conditions for 48 h, were shown to be stable with no significant change over this period (Table 3). These results are indicated (0.5% changes in area between $T = 0$ h and $T = 48$ h). Based on these data that show quantitative recovery through 48 h, solutions of methyl salicylate can be assayed within 48 h of preparation.

TABLE 3 Stability data for methyl salicylate

Time (h)	RT (min)*	Area (mAU s)	Height (mAU)	Recovery (%)
0	0.09	0.18	0.26	99.98
48	0.12	0.22	0.31	98.96

*RSD (%, $n = 3$)

3.2 Method Validation

3.2.1 Linearity

Linearity was studied using seven solutions in the concentration range 25-175 µg/mL. Solutions corresponding to each concentration level were injected in triplicate and linear regression analysis of the methyl salicylate peak area (*y*) versus methyl salicylate concentration (*x*) was calculated. The correlation coefficient (r^2 = 0.9999) obtained for the regression line demonstrates that there is a strong linear relationship between peak area and concentration of methyl salicylate (Table 4).

TABLE 4 Method validation results of methyl salicylate

Validation step	Parameters	Conc. (µg/mL)	Results	Acceptance criteria
Linearity	(k = 7, n = 3)	25-175	y = 1.4748x + 7788.3 (r^2 = 0.9999)	$r^2 \geq 0.999$
Repeatability (Area)	RSD (%, n = 10)	100	0.14	X < 2
Intermediate precision				
Day 1, LC 1, analyst 1	RSD (%, n = 6)	100	0.19	X < 2
Day 2, LC 2, analyst 2	RSD (%, n = 6)	100	0.22	X < 2

3.2.2 Accuracy/recovery study

Accuracy of the method was evaluated by fortifying a methyl salicylate sample solution (200 µg/mL) with three known concentrations of reference standard (50, 100, and 150 µg/mL). Percent recoveries were calculated form differences between the peak areas obtained for fortified and unfortified solutions. Good recoveries were obtained (Table 5). No significant differences were observed between amounts of methyl salicylate added and the amounts found ($p < 0.05$).

TABLE 5 Recovery studies of methyl salicylate from samples with known concentration

Sample #	Percent of nominal	Amount of analyte (µg/mL)		Recovery (%, $n = 3$)	RSD (%, $n = 3$)
		Added	Found		
1	50	5.0	4.98	99.6	0.09
2	100	9.0	8.98	99.7	0.12
3	150	13.0	12.99	99.9	0.15

3.2.3 Precision

The precision of the method was determined by repeatability (intra-day) and intermediate precision (inter-day variation). Repeatability was examined by analysing six determinations of the same batch of each component at 100% of the test concentration. The RSD of the areas of methyl salicylate peak were found to be less than 0.14% (Table 4), which confirms that the method is sufficiently precise. Intermediate precision (inter-day variation) was studied by assaying five samples containing the nominal amount of methyl salicylate on different days by different analysts using different LC systems. Solutions corresponding to each concentration level were injected in duplicate. The RSD values across the systems and analysts were calculated and found to be less than 0.23% (Table 4) for each of the multiple sample preparation, which demonstrates excellent precision for the method.

3.2.4 Specificity

The LC-PDA/UV isoplot chromatogram (Figure 5) demonstrates a good separation of the methyl salicylate. The isoplot chromatogram data consist of UV absorption spectra from 200 to 400 nm for each point along the chromatogram. Injections of the extracted placebo were performed to demonstrate the absence of interference with the elution of the methyl salicylate. This result demonstrates (Figure 6) that there was no interference from the other materials in the cream formulation and, therefore, confirms the specificity of the method.

Forced degradation studies were performed to evaluate the specificity of methyl salicylate under four stress conditions (heat, UV light, acid, base). Solutions of methyl salicylate were exposed to 60°C for 1 h, UV light using a UVL-56 lamp for 24 h, acid (1 M hydrochloric acid) for 24 h and base (1 M sodium hydroxide) for 4 h. A summary of the stress results for retention time (TR), peak area, area percent, resolution (R_s) and column efficiency (theoretical plate numbers) is shown in Table 6. Under acid (major degradation) hydrolysis condition, the methyl salicylate content decreased and additional peak was observed (Figure 7). No degradation was observed under other hydrolysis conditions (heat, UV light and base) studied (Figure 7). The additional peak detected at 1.83 min under acid condition. This was further confirmed by peak purity analysis on a PDA UV detector. The methyl salicylate analyte obtained by acid hydrolysis was well resolved (5.82 min) from the additional peak (Figure 6), indicating the specificity of the method.

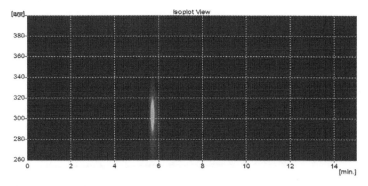

FIGURE 5 HPLC PDA/UV isoplot chromatogram of methyl salicylate.

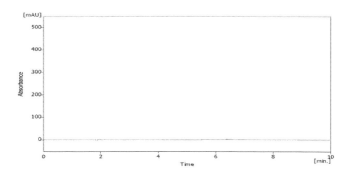

FIGURE 6 HPLC chromatogram of placebo.

TABLE 6 Force degradation studies data for methyl salicylate

Stress conditions	Sample treatment	RT (min)	Area (mAU.s)	Assay (%)	R_s	Efficiency (th.pl)
Reference	Fresh solution	5.82	6234.93	99.72	-	4686
Heat	60 °C for 1 h	5.80	6364.56	98.94	-	4659
Light	UV Light for 24 h	5.82	6319.64	98.87	-	4686
Acid	1M HCl for 24 h				5.3	
		5.82	5354.81	75.6	22	4686
		1.83*	1637.44	24.4	-	4190
Base	1M NaOH for 4 h	5.80	4895.87	98.95		5545

*Degradation peak

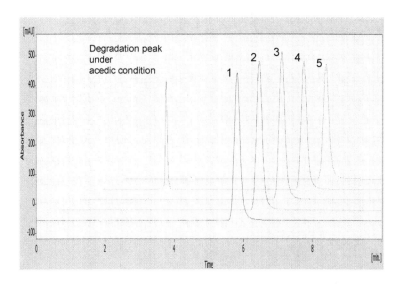

FIGURE 7 HPLC chromatograms of methyl salicylate obtained under stress conditions (1) fresh reference standard; (2) heat at 60°C; (3) UV light; (4) acid; (5) base.

3.2.5   Limits of detection and quantitation

The limit of detection (LOD) and limit of quantitation (LOQ) of methyl salicylate was determined based on standard deviation (σ) of response and slope (s) (10). Methyl salicylate solutions were prepared in the range 0.02-25 µg/mL and injected in triplicate. Average peak area of analyte was plotted against concentration. LOD and LOQ were calculated by using the following equations:

LOD = (3.3 σ)/s   (Eq.1)

LOQ = (10 σ)/s   (Eq.2)

The LOD was determined to be 2.4 µg/mL and LOQ was found to be 14 µg/mL for methyl salicylate with CV less than 2% for six replicate injections.

4. CONCLUSION

A new reversed-phase liquid chromatographic method with UV spectrophotometric detection was developed for the determination of methyl salicylate in medicated cream formulation. The method was validated and the results obtained were accurate and precise with RSD < 1% in all cases and no significant interfering peaks were detected. The method is specific, simple, selective, robust and reliable for routine use in quality control analysis of methyl salicylate raw materials, bulk samples and final medicated cream product release.

REFERENCES

1. National Research Council, Assessment of the U.S. Army Chemical and Biological Defense Command, Report 1, Technical Assessment of the Man-In-Simulant Test (MIST) Program, pp. 30-36, National Academy Press, Washington, DC, 1997.
2. Barker, R.L. A review of gaps and limitations, in test method for first responder protective clothing and equipment, final report presented to the National Personal Protection Technology Laboratory at the National Institute for occupational safety and Health, p. 37, 2005.
3. Muscle-Pain Reliever Is Blamed for Staten Island Runner's Death". New York Times, 2007-06-10.

4. Henry, S.I. Tan, P.A.K., Petra, E.P. Gas-liquid chromatographic assay of mixtures of camphor, menthol, and methyl salicylate in ointments. *J. Chromatogr. A* 238(1), 241- 246, 1982.

5. Shabir, G.A., Lough, W.J., Shafique, A.A., Bradshaw, T.K. Evaluation and application of best practice in analytical method validation. *J. Liq. Chromatogr. Relat. Technol.* 30(3), 311-333, 2007.

6. Shabir, G.A. Validation of HPLC methods for pharmaceutical analysis: Understanding the differences and similarities between validation requirements of the U.S. FDA, USP and ICH. *J. Chromatogr. A,* 987(1-2), 57- 66, 2003.

7. Shabir, G.A. Step-by-step analytical methods and protocol in the quality system compliance industry. *J. Validation Technol.* 10(4), 314-324, 2004.

8. Shabir, G. A. A practical approach to validation of HPLC methods under current good manufacturing practices. *J. Validation Technol.* 10(3), 210-218, 2004.

9. Shabir, G. A. HPLC method development and validation for pharmaceutical analysis. *Pharma. Technol. Eur.* 16(3), 37-49, 2004.

10. International Conference on Harmonization (ICH). Validation of analytical procedures: Text and Methodology, Q2 (R1), Switzerland, Geneva, 2005.

11. Reviewer Guidance: Validation of Chromatographic Methods, FDA Center for Drug Evaluation and Research (CDER), NJUnited States, 1994.

12. U.S. Pharmacopeia 34, Chromatography, General Chapter (621), United States Pharmacopeal Convention, Rockville Maryland, p. 2011.

11 DETERMINATION OF 1,7,7-TRIMETHYL BICYCLO(2,2,1) HEPTAN-2-ONE IN A CREAM PHARMACEUTICAL FORMULATION BY HPLC

ABSTRACT

A new reversed-phase high-performance liquid chromatographic (HPLC) method for the determination of 1,7,7-trimethyl-bicyclo(2,2,1)heptan-2-one in a cream formulation is developed and validated. The separation was achieved using an isocratic mobile phase, on a Lichrosorb C_8 column. The calibration curve is linear (r^2 = 0.9999) from 25-175% of the analytical concentration of 1 mg/mL. The mean percent standard deviation values for intra-day and inter-day precision studies were <1%. The recovery ranges 99.80-100.06% from a cream formulation. The method was found to be precise, accurate, specific and robust during the study. The method can be used reliably in quality control for the analysis of bulk cream samples and final product release.

Keywords: Reversed-phase HPLC, Method validation, 1,7,7-Trimethyl-bicyclo(2,2,1)heptan-2-one, Pharmaceutical cream formulation

1. INTRODUCTION

1,7,7-Trimethyl-bicyclo(2,2,1)heptan-2-one (Figure 1) is used for its scent, as an embalming fluid and in pharmaceutical formulations such as in antiseptic and anti-itching creams as the active ingredient. It is a white transparent waxy crystalline solid with a strong penetrating pungent aromatic odour [1]. It is found in wood of the 1,7,7-trimethyl-bicyclo(2,2,1)heptan-2-one laurel, Cinnamonum camphora, which is a large evergreen tree found in Asia (particularly in Borneo, hence its alternate name); it can also be synthetically produced from oil of turpentine. Turpentine (also called spirit of turpentine, oil of turpentine, wood turpentine, gum turpentine) is a fluid obtained by the distillation of resin obtained from trees, mainly pine trees. It is composed of terpenes, mainly the monoterpenes alpha-pinene and beta-pinene [2]. 1,7,7-Trimethyl-bicyclo(2,2,1)heptan-2-one is readily absorbed through the skin and produces a feeling of cooling and acts as a slight local anesthetic and antimicrobial agent. Recently, carbon nanotubes were successfully synthesized using 1,7,7-trimethyl-bicyclo(2,2,1)heptan-2-one in chemical vapor deposition process [3]. In larger quantities, it is poisonous when ingested and can cause seizures, confusion, irritability, and neuromuscular hyperactivity. Despite the rather low skin absorption it may still lead to hepatotoxicity in extreme cases [4,5]. In 1980, the United States Food and Drug Administration (FDA) set a limit of 11% allowable camphor in consumer products and totally banned products labeled as camphorated oil, camphor oil, camphor liniment, and camphorated liniment (except "white camphor essential oil", which contains no significant amount of camphor). Since alternative treatments exist, medicinal use of 1,7,7-trimethyl-bicyclo(2,2,1)heptan-2-one is discouraged by the FDA, except for skin-related uses, such as medicated powders, which contain only small amounts of1,7,7-trimethyl-bicyclo(2,2,1)heptan-2-one. Various analytical procedures for analysis of 1,7,7-trimethyl-bicyclo(2,2,1)heptan-2-one with the combination of other compounds including *m*-cresols [6], menthol [7] and mixtures of1,7,7-trimethyl-bicyclo(2,2,1)heptan-2-one, menthol, and methyl salicylate in ointments [8] using gas chromatography have been reported in the literature. In this work for the first time, a simple, rapid, specific and sensitive reversed-phase high-performance liquid chromatography (RP-HPLC) novel assay method for the determination of 1,7,7-trimethyl-bicyclo(2,2,1)heptan-2-one in a pharmaceutical cream formulation is reported. Analytical method development and validation is an important part of analytical chemistry and plays a major role in the discovery, development, and manufacture of pharmaceuticals. The official test methods that result from these processes are used by quality control laboratories to ensure the identity, purity, potency and performance of drug product 'quality' essential for drug safety and efficacy. Frequently, RP-HPLC is the analytical method of choice in pharmaceutical analysis because of its specificity (i.e. all the components of a sample are separated from one another before the measurement is made so that its results arise from the analyte and from nothing else). Finally, the developed analytical method was validated to assess the validity of research data means determining whether the method used during the study can be trusted to provide a genuine, account of the intervention being evaluated. As a best practice [9-13], in the subsequent investigation, the new RP-HPLC method was validated [14] using pre-approved protocol and validated HPLC system.

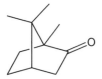

FIGURE 1 Structure of 1,7,7-trimethyl-bicyclo(2,2,1)heptan-2-one.

2. EXPERIMENTAL

2.1 Chemicals and Reagents

Methanol (HPLC-grade), 1,7,7-trimethyl-bicyclo(2,2,1)heptan-2-one (96% pure), ammonium formate, and ammonium acetate were obtained from Sigma-Aldrich (Gillingham, UK). Acetic acid (analytical grade) was purchased from BDH (Lutterworth, UK). Distilled water was de-ionised by using a Milli-Q system (Millipore, Bedford, MA).

2.2 HPLC Instrumentation and Conditions

A Knauer (Berlin, Germany) HPLC system, consisting of a vacuum degasser, a Knauer 1000 solvent pump, Knauer 3950 autosampler and a Knauer photodiode-array (PDA) detector 2600, all controlled by a ClarityChrom software, was used. A reversed phase Lichrosorb C_8 column (150 mm × 4.6 mm, 5 µm particle size) (Jones

Chromatography, Hengoed, UK) was used for separation. The mobile phase comprised of a (65:35, v/v) mixture of methanol-water containing 0.1% analytical grade acetic acid (v/v) was used. The flow rate was 1 mL/min, the injection volume was 20 μL, and the temperature was set at 30°C ±0.5°C.

2.3 Standards Preparation

1,7,7-Trimethyl-bicyclo(2,2,1)heptan-2-one (2.5 g) was accurately weighed and added to a 100 mL volumetric flask before being dissolved in methanol (stock). A 4.0 mL aliquot of stock solution was diluted to 100 mL in mobile phase, yielding a final concentration of 1 mg/mL. Standard solutions for the evaluation of 1,7,7-trimethyl-bicyclo(2,2,1)heptan-2-one linearity were prepared over a concentration range of 0.25-1.75 mg/mL, to 25, 50, 75, 100, 125, 150, and 175%, in the mobile phase.

2.4 Sample Preparation

An accurately weighed amount (10.0 g) of sample cream into a 150 mL stopper quick fit flat bottomed flask and 36 mL methanol and 1 mL acetic acid were added. The flask was heated on a boiling water bath until the sample cream melted. The solution was then cooled to room temperature. Forty milliliter of de-ionized water was added to the flask and again sample was heated to melt the cream and then cooled to room temperature. Amount equivalent to 100 mg 1,7,7-trimethyl-bicyclo(2,2,1)heptan-2-one from pharmaceutical formulation was transferred to 100 mL volumetric flask and 30 ml mobile phase was added, yielding a final concentration of 1 mg/mL. The flask was shaken mechanically for 10 min and the solution was then diluted to volume with mobile phase. The sample was filtered through a sample filtration unit (0.45 μm) and injected into the HPLC system.

3. RESULTS AND DISCUSSION

3.1 Method Development

Acetic acid in water, ammonium formate, and ammonium acetate buffers were studied as potential mobile phases but acetic acid was selected because it gave a shorter analysis time and better separation was obtained with the addition of methanol. Initially, three analytical columns were tried in order to reach an acceptable specificity and selectivity for camphor compound. We first exploited Hypersil C_{18} (250 mm × 4.60 mm, 5μm) and HyperClone C_{18} (250 mm × 4.60 mm, 5μm) phase columns from Phenomenex (Macclesfield, UK). The Hypersil column gave poor separation of camphor peak with retention time of 5.35 min (Figure 2). The Hyperclone column also gave a poor peak shape with tailed badly and was longer retention time 12.58 min (Figure 3) for the analyte peak. Shift to the Lichrosorb C_8 (150 mm × 4.6 mm, 5 μm), column (Jones Chromatography, Hengoed, UK) produced a peak with superior band shape and a column efficiency with shorter retention time (5.97 min) under the same conditions (Figure 4). The temperature was set at 30°C ±0.5°C. The choice of wavelength is essential to accomplish a sensitive chromatographic assay. The optimal wavelength for 1,7,7-trimethyl-bicyclo(2,2,1)heptan-2-one detection was established using the scan range of 200 to 400 nm. It was established that 289 nm was the optimal wavelength to maximize the signal.

System suitability testing was performed by injecting six replicate injections of a solution containing 1 mg 1,7,7-trimethyl-bicyclo(2,2,1)heptan-2-one/mL. The percent relative standard deviation (RSD) of the peak area responses was measured, giving an average of 0.11 (n = 6). The tailing factor (T) for each 1,7,7-trimethyl-bicyclo(2,2,1)heptan-2-one peak was 1.071, the theoretical plate number (N) was 5615, and the retention time (t_R) variation %RSD was < 1% for six injections. The RPLC method met these requirements within the accepted limits.[15] For the determination of method robustness within a laboratory during method development a number of chromatographic parameters were determined, which included flow rate, temperature, mobile phase composition, and columns from different lots, In all cases good separation of 1,7,7-trimethyl-bicyclo(2,2,1)heptan-2-one were always achieved, indicating that the method remained selective for 1,7,7-trimethyl-bicyclo(2,2,1)heptan-2-one component under the tested conditions.

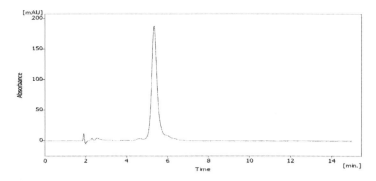

FIGURE 2 Chromatogram of 1,7,7-trimethyl-bicyclo(2,2,1)heptan-2-one obtained using a Hypersil C_{18} (250 mm x 4.60 mm, 5μm particle size) column.

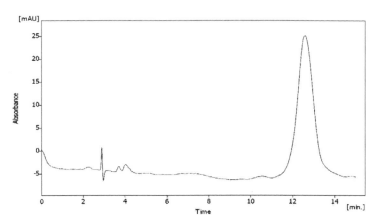

FIGURE 3 Chromatogram of 1,7,7-trimethyl-bicyclo(2,2,1)heptan-2-one obtained using HyperClone C_{18} (250 mm x 4.60 mm, 5μm particle size) column.

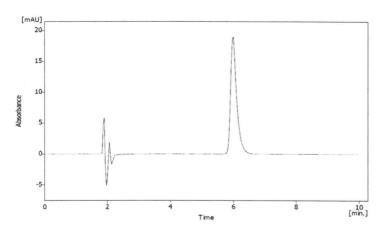

FIGURE 4 Chromatogram of 1,7,7-trimethyl-bicyclo(2,2,1)heptan-2-one obtained using Lichrosorb C_8 (150 mm x 4.6 mm, 5 µm particle size) column.

To determine the robustness of method, the final experimental conditions were purposely altered and the results were examined. The flow rate was varied by 1 ±0.2 mL/min. the percentage of organic modifier was varied by 65 ± 5% and temperature was varied by 30 ±5°C. Their effects on the retention time (t_R), tailing factor (T), theoretical plate numbers (N) and repeatability of peak areas (n = 3) were studied. It can be seen that every employed condition, the chromatographic parameters are in accordance with established value [15, 16]. A change of mobile phase composition, flow rate and temperature had no impact on chromatographic performance. The tailing factor for 1,7,7-trimethyl-bicyclo(2,2,1)heptan-2-one peak was found to be less than 1.2 and analyte was well separated under all the changes carried out (Table 1). Considering the result of modifications in the system suitability parameters and the specificity of the method, it would be concluded that the method conditions are robust.

TABLE 1 Robustness data of HPLC method for 1,7,7-trimethyl-bicyclo(2,2,1)heptan-2-one

Parameter	Modification	TR (min)	T	N	RSD (%)*
Mobile phase	60%	6.405	1.018	5612	0.16
(Methanol, ± 5%)	65%	6.400	1.018	5687	0.11
	70%	6.398	1.015	5567	0.13
Flow rate (± 0.2,	0.8	6.408	1.018	5689	0.11
mL/min)	1.0	6.400	1.018	5685	0.16
	1.2	6.397	1.018	5681	0.11
Temperature (±	25	6.400	1.018	5688	0.14
5°C)	30	6.400	1.019	5687	0.12
	35	6.399	1.018	5653	0.13

*Peak area ($n = 5$)

3.2 Method Validation

3.2.1 Stability of analytical solutions

Samples and standard solutions were chromatographed immediately after preparation and then reassyed after storage at room temperature for 48 h. The results given in Table 1 show that there was no significant change (<1% response factor) in 1,7,7-trimethyl-bicyclo(2,2,1)heptan-2-one concentration over this period.

3.2.2 Linearity

Linearity was studied using seven solutions in the concentration range 0.25-1.75 mg/mL, 25-175% around the theoretical value (1 mg/mL). Solutions corresponding to each concentration level were injected in triplicate and linear regression analysis of the 1,7,7-

trimethyl-bicyclo(2,2,1)heptan-2-one peak area (*y*) versus 1,7,7-trimethyl-bicyclo(2,2,1)heptan-2-one concentration (*x*) was calculated. The correlation coefficient (r^2 = 0.9999) obtained for the regression line demonstrates that there is a strong linear relationship between peak area and concentration of 1,7,7-trimethyl-bicyclo(2,2,1)heptan-2-one (Table 2).

TABLE 2 Method validation results

Validation step	Parameters	Conc (mg/mL)	Results	Acceptance criteria
Standard stability	% change in response factor	1.0	0.12	X < 2
Sample stability	% change in response factor	1.0	0.14	X < 2
Linearity (*n* = 3)	Correlation coefficient (r^2)	0.25-1.75	y = 725.18x - 91.577 (r^2 = 0.9999)	r^2 = ≥ 0.999
Repeatability (*n* = 6)	t_R (min) %RSD	1.0	0.04	X < 2
	Peak area %RSD		0.16	X < 2
Intermediate precision (*n* = 3)	Instrument %RSD	1.0	0.26	X < 2
	Analyst %RSD		0.18	X < 2

3.2.3 Precision

The precision of the method was investigated with respect to repeatability (intra-day precision) and intermediate precision (inter-day variation). Repeatability of the method was evaluated by assaying six replicate injections of the 1,7,7-trimethyl-bicyclo(2,2,1)heptan-2-one at 100% of test concentration (1 mg/mL). The %RSD of the retention time (min) and relative percent peak area were found to be less than 0.17% (Table 2). Intermediate precision (inter-day variation) was demonstrated by two analysts using two HPLC systems and evaluating the relative peak area percent data across the two HPLC systems at three concentration levels (50, 100, and 125%) that cover the assay method range (0.25-1.75 mg/mL). the mean and %RSD across the systems and analysts were calculated from the individual relative percent peak area mean values at the 50%, 100%, and 125% of the test concentration. The %RSD values for both instruments and analysts were $\leq 0.26\%$ (Table 2) and illustrated good precision of RP-HPLC method.

3.2.4 Accuracy/recovery studies

The accuracy of the method was evaluated by adding known quantities of 1,7,7-trimethyl-bicyclo(2,2,1)heptan-2-one in the cream formulation samples to give a range of 1,7,7-trimethyl-bicyclo(2,2,1)heptan-2-one concentration of 75-150% ($n = 3$) of that in a test preparation. These solutions were analyzed and the amount of analyte recovered calculated. The recovery data expressed as an average percent of triplicate injections are presented in Table 3 and show good recovery of 1,7,7-trimethyl-bicyclo(2,2,1)heptan-2-one.

TABLE 3 Recovery studies of 1,7,7-trimethyl-bicyclo(2,2,1)heptan-2-one from samples with known concentration

Sample #	Percent of nominal	Amount of analyte (mg)		Recovery (%) ($n = 3$)	%RSD ($n = 3$)
		Added	Recovered		
1	75	0.75	0.7502	100.02	0.12
2	100	1.0	0.998	99.80	0.17
3	150	1.50	1.501	100.06	0.14
Mean				99.96	

3.2.5 Specificity/selectivity

Injections of the extracted placebo were performed to demonstrate the absence of interference with the elution of the 1,7,7-trimethyl-bicyclo(2,2,1)heptan-2-one.these results demonstrate (Figure 5) that there was no interference from the other materials in the cream formulation and, therefore confirm the specificity of the RP-HPLC method.

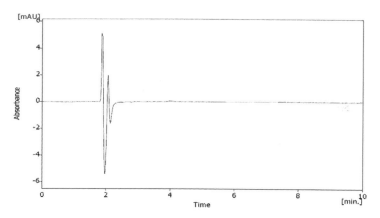

FIGURE 5 Chromatogram of placebo without adding active analyte in the formulation.

The forced degradation studies were also applied to 1,7,7-trimethyl-bicyclo(2,2,1)heptan-2-one reference standard at a concentration of 1 mg/mL to verify that none of the degradation products interfered with quantitation of the drug. Hydrolytic degradation was studied by heating the drug under reflux at 80°C in 0.1 M hydrochloric acid and 0.1 M sodium hydroxide for 4 h. The samples were then cooled to room temperature and neutralized. Oxidative degradation was studied by treating the drug with 3% hydrogen peroxide at room temperature (22 ±1°C) for 4 h. Solutions containing 1 mg/mL of each degraded samples were prepared and injected in triplicate. No significant degradation was observed under any stress conditions studied. (Table 4)

TABLE 4 Results of the stress conditions of 1,7,7-trimethyl-bicyclo(2,2,1)heptan-2-one experiments.

Stress conditions	Sample treatment	t_R (min)	Area (mAU s)	Assay (%)
Reference	Fresh solution	6.400	7921	99.98
Acid	1M HCl at 80°C for 4 h	6.398	7916	99.86
Base	1M NaOH at 80°C for 4 h	6.397	7922	99.77
Oxidative	3% H_2O_2 at room temperature for 24 h	6.400	7911	99.68

3.2.6 Limits of detection and quantitation

The limit of detection (LOD) and limit of quantitation (LOQ) of 1,7,7-trimethyl-bicyclo(2,2,1)heptan-2-one was determined based on standard deviation (σ) of response and slope (s) [14]. 1,7,7-Trimethyl-bicyclo(2,2,1)heptan-2-one solutions were prepared in the range 5-250 µg/mL and injected in triplicate. Average peak area of analyte was plotted against concentration. LOD and LOQ were calculated by using the following equations:

LOD = (3.3 σ)/s   (Eq.1)

LOQ = (10 σ)/s   (Eq.2)

The LOD was determined to be 8.4 µg/mL and LOQ was found to be 76 µg/mL for 1,7,7-trimethyl-bicyclo(2,2,1)heptan-2-one with %RSD less than 0.16% for six replicate injections.

4. CONCLUSION

A new reversed-phase HPLC method with UV spectrophotometric detection was developed for the determination of 1,7,7-trimethyl-bicyclo(2,2,1)heptan-2-one in pharmaceutical cream formulation. The method was validated and the results obtained were accurate and precise with RSD < 1% in all cases and no significant interfering peaks were detected. The method is specific, simple, selective, robust and reliable for routine use in quality control for analysis of 1,7,7-trimethyl-bicyclo(2,2,1)heptan-2-one in bulk, raw materials, and final cream products pharmaceutical formulations.

REFERENCES

1. Mann, C.; Hobbs, J.B.; Banthorpe, D.V.; Harborne, J.B. Natural products: their chemistry and biological significance. Harlow, Essex, England: Longman Scientific & Technical, 1994, pp. 309-311.
2. Barnhart, R.K. The barnhart consise dictionary of etymology. Harper Collins, New York, 1995.
3. Kumar, M.; Ando, Y. Carbon nanotubes from camphor: An environment-friendly nanotechnolog. *J. Phys. Conf. Ser.* 2007, 61, 643–646.
4. Martin, D.; Valdez, J.; Boren, J.; Mayersohn, M. Dermal absorption of camphor, menthol, and methyl salicylate in humans. *J. Clin. Pharmacol.* 2004, 44, 1151-1157.
5. Uc, A.; Bishop, W.P.; Sanders, K.D. Camphor hepatotoxicity. *South Med. J.* 2000, 93, 596-598.
6. Mirza, T.; Tan, S.I.H. Capillary gas chromatographic assay of camphor and *m*-cresol in dermatological creams. *J. Pharm. Biomed. Anal.* 1998, 17(8), 1427-1438.

7. Karuza, L.J.; Folivarski, K. Validation of the assay method for camphor and menthol in a herbal drug preparation. *J. Pharm. Biomed. Anal.* 1996, 15(3), 419-422.

8. Henry, S.I.T.; Patricia, A.K.; Petra, E.P. Gas-liquid chromatographic assay of mixtures of camphor, menthol, and methyl salicylate in ointments, *J. Chromatogr. A* 1982, 238(1), 241-246.

9. Shabir, G.A.; Lough, W.J.; Shafique, A.A.; Bradshaw, T.K. Evaluation and application of best practice in analytical method validation. *J. Liq. Chromatogr. Rel. Technol.* 2007, 30 (3), 311-333.

10. Shabir, G.A. Validation of HPLC methods for pharmaceutical analysis: Understanding the differences and similarities between validation requirements of the U.S. Food and Drug Administration, the U.S. Pharmacopoeia and the International Conference on Harmonization. *J. Chromatogr. A.* 2003, 987(1-2),   57-66.

11. Shabir, G.A. Step-by-step analytical methods and protocol in the quality system compliance industry. *J. Validation Technol.* 2004, 10(4), 314-324.

12. Shabir, G. A. A practical approach to validation of HPLC methods under current good manufacturing practices. *J. Validation Technol.* 2004, 10(3), 210-218.

13. Shabir, G. A. HPLC method development and validation for pharmaceutical analysis. *Pharma. Technol. Eur,* 2004, 16(3), 37-49.

14. International Conference on Harmonization (ICH). Validation of analytical procedures: Text and Methodology, Q2(R1), Switzerland, Geneva, 2005.

15. U.S. Pharmacopeia 32, Chromatography, General Chapter (621), United States Pharmacopeal Convention, Rockville Maryland, 2009, p. 1776.

16. Reviewer Guidance: Validation of Chromatographic Methods, Food and Drug Administration (FDA) Center for Drug Evaluation and Research (CDER), United States, 1994.

12 DETERMINATION OF PARACETAMOL, METHYL PARAHYDROXYBENZOATE, ETHYL PARAHYDROXYBENZOATE AND PROPYL PARAHYDROXYBENZOATE IN MEDICINAL SUSPENSION AND TABLETS BY REVERSED-PHASE HPLC/UV-DAD

ABSTRACT

This paper presents a new reversed-phase high-performance liquid chromatographic (HPLC) method with ultraviolet absorbance diode-array detection (UV-DAD) for simultaneous determination of paracetamol, methyl parahydroxybenzoate, propyl parahydroxybenzoate and ethyl parahydroxybenzoate in a commercial medicinal oral suspension and tablets. The chromatographic separation is achieved with acetonitrile, tetrahydrofuran and water in ratio of 21:13:66 (v/v/v) adjusted to pH 3 with formic acid as mobile phase, a Lichrosorb RP8 (150 mm × 4.6 mm, 5 μm) column and UV detection at 249 nm. The linear range of determination for paracetamol, methyl parahydroxybenzoate, propyl parahydroxybenzoate and ethyl parahydroxybenzoate are 25-160 μg/mL, 20-128 μg/mL, 5-40 μg/mL and 5-32 μg/mL, respectively. The correlation coefficients are ≥ 0.9997 for each analyte. The mean percent relative standard deviation values for intra- and interday precision studies are < 1% in each case. The applicability of the method was evaluated in commercial dosage forms analysis.

Keyword: HPLC/UV-DAD, Paracetamol, Methyl parahydroxybenzoate, Propyl parahydroxybenzoate, Ethyl parahydroxybenzoate, Oral suspension, Tablets

1. INTRODUCTION

Paracetamol (acetaminophen) is a widely used over-the-counter analgesic and antipyretic drugs. As pain and fever are common, no home should be without some paracetamol, particularly homes with children. Paracetamol (PCM) is available in many different pharmaceutical preparations such as tablets, capsules and liquid suspensions. PCM is used as an active ingredient with combination of preservatives namely methyl parahydroxybenzoate (MHB), ethyl parahydroxybenzoate (EHB) and propyl parahydroxybenzoate (PHB) in pharmaceutical oral suspension formulation. These preservatives have been widely used as antimicrobial and anti-fungal agents as single preservative but more often in combinations in food, beverage, cosmetics and pharmaceutical formulations [1,2]. These preservatives are commonly added to these formulations due to their broad antimicrobial spectrum with good stability and non-volatility [2] to prevent alteration and degradation of the product formulations and to prolong product shelf life. Industry must be fully aware of the procedure for preservative systems in a product need to be analysed to establish their effectiveness throughout shelf life of the product. Hence, the simultaneous determination of these preservatives in commercial pharmaceutical products is particularly important both for quality assurance and consumer safety.

Many existing analytical procedures are available in literature for the determination of MHB, EHB and PHB, either alone or in combination with other drugs by high-performance liquid chromatography (HPLC) and other techniques [2-15]. Additionally, different analytical methods are available for quantification of paracetamol [16,17]. The novelty of the proposed method is the simultaneous analysis of PCM drug and a combination of the three preservatives in pharmaceutical dosage forms.

The aim of this study was to develop a new, simple, rapid and robust HPLC method with ultraviolet diode-array detection (UV-DAD) for the simultaneous determination of PCM, MHB, EHB and PHB compounds in a single chromatographic run from pharmaceutical dosage forms.

As a best practice [18,19] in the subsequent investigation, the new HPLC with diode-array detection assay method was validated according to criteria described in the literature [20-24] and its applicability was evaluated in commercial pharmaceutical dosage forms.

2. EXPERIMENTAL

2.1 Materials
Acetonitrile and methanol (HPLC-grade) were obtained from Fisher Scientific (Loughborough, UK). HPLC-grade tetrahydrofuran (THF)

was obtained from BDH (Poole, UK). Acetaminophen (purity ≥ 99%), methyl parahydroxybenzoate (purity ≥ 99%), ethyl parahydroxybenzoate (purity ≥ 99%), propyl parahydroxybenzoate (purity ≥ 99%) and formic acid were purchased from Sigma-Aldrich (Gillingham, UK). Paracetamol oral suspension and paracetamol tablets were purchased from local pharmacy (Oxford, UK). Ultra-purified (deionised) water was prepared in-house using a Millipore Milli-Q water system (Watford, UK).

2.2 HPLC Instrumentation and Conditions

A Knauer HPLC system (Berlin, Germany) equipped with a model 1000 LC pump, an online degasser, model 3950 autosampler, and model 2600 ultraviolet diode-array detector was used. The data were acquired via Knauer ClarityChrom Workstation data acquisition software. The mobile phase consisted of a mixture of acetonitrile, tetrahydrofuran and water in ratio of 21:13:66 (v/v/v) adjusted to pH 3.0 with formic acid was used, at a flow rate of 1 mL/min. The injection volume was 10 μL and the detection wavelength was set at 249 nm. RPLC analysis was performed isocratically at ambient temperature using Jones Chromatography column Lichrosorb RP8 (150 × 4.6 mm, 5 μm particle).

2.3 Standard Preparation

A combined standard stock solution of PCM drug and MHB, EHB preservatives was prepared in mobile phase, yielding a final concentration of 0.1, 0.08, 0.025 and 0.02 mg/mL, respectively.

2.4 Sample Preparation

An accurately weighed amount (5 g) of PCM suspension sample was placed in a 100 mL volumetric flask and dissolved in methanol. Five millilitre aliquot solutions was transferred to a 50 mL volumetric flask and diluted in mobile phase, yielding a final concentration of 0.12 mg/mL.

The mean weight of finally powdered PCM tablets containing 500 mg of PCM was accurately transferred into 100 mL volumetric flask and about 70 mL of methanol was added; the mixture was extracted in the ultrasonic bath for 10 min at room temperature and diluted with mobile phase to the mark. The supernatant liquid was filtered through 0.22 μm filter. Five millilitre of this solution was transferred to the 50 mL volumetric flask and diluted with mobile phase to the mark, yielding a final concentration of 0.1 mg/mL.

3. RESULTS AND DISCUSSION

3.1 Method Development

The procedure for the simultaneous determination of PCM drug and MHB, EHB, PHB preservatives using reversed phase HPLC/UV-DAD is reported. The mobile phase was chosen after several trials with methanol, acetonitrile, tetrahydrofuran, water and buffer solutions in various compositions and at different pH values. The best separation was obtained using the mobile phase consisted of a mixture of acetonitrile, tetrahydrofuran and water in ratio of 21:13:66 (v/v/v). Flow rates between 0.5 and 1.5 mL/min were studied. A flow rate of 1 mL/min gave an optimal signal to noise ratio with excellent separation time. The photodiode-array detector was set at 200 to 400 nm and PCM, MHB, EHB and PHB components were extracted at maximum absorption at 249 nm and this wavelength was chosen for the assay method.

Preliminary experiments were performed to select the column most suitable for the separation of the analytes consider the physical and chemical properties of the analyte(s), the mode of analysis and how the analytes will intract with the surface of the chromatographic phase. A C18 and RP8 column were tried in the following order: Nucleosil C18 (25 cm × 4.6 mm, 10µm) and Lichrosorb RP8 (15 cm × 4.6 mm, 5 µm particle). The C18 column gave a poor separation, even when changing the composition of the mobile phase. Lichrosorb RP8 gave satisfactory results under the same experimental conditions (see HPLC Instrumentation and Conditions section). Under these conditions, all analytes were eluted in less than 15 min with acceptable separation. Using Lichrosorb RP8, the retention times for PCM, MHB, EHB and PHB were found to be 2.38, 5.48, 7.98 and 12.70 min, respectively (Figure 1a). For the determination of method robustness, a number of chromatographic parameters were determined, which included flow rate, temperature, mobile phase composition, wavelength accuracy and column from different lots. In all cases, good separations of PCM drug and MHB, EHB, PHB preservatives were always achieved, indicating that the method remained selective for PCM drug and MHB, EHB, PHB preservatives under the tested conditions.

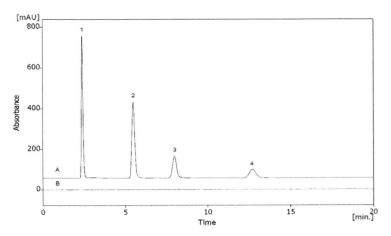

FIGURE 1 Overlay HPLC chromatogram obtained from sample suspension (a) PCM (peak 1, tR = 2.38), MHB (peak 2, tR = 5.48), EHB (peak 3, tR = 7.98) and PHB (peak 4, tR = 12.70 min), (b) placebo without adding analytes in the formulation.

System suitability testing verifies that the HPLC system is working as expected. It is based on the concept that the equipment, electronics, analytical operations, and samples to be analysed constitute an integral system. System suitability was evaluated by injecting mixture of standard solutions containing 0.1, 0.08, 0.025 and 0.02 mg/PCM, MHB, EHB and PHB/mL in replicates of six at the beginning of validation run. All system suitability acceptance criteria were met. Summaries of all the system suitability results are given in Table 1.

TABLE 1 System suitability parameters for the HPLC method

Parameters	Acceptance Limits	Results PCM	MHB	EHB	PHB
Retention time (min)	-	2.60	5.93	8.68	13.77
Resolution (Rs)	$R_s > 2$	-	11.55	6.10	7.59
Capacity factor (K')	$K' > 2$	2.38	6.70	10.27	16.88
Tailing factor (T)	$T \leq 2$	1.24	1.20	1.06	1.08
Theoretical plates (N)	$N > 2000$	2476	4447	4642	4986
Injection precision ($n = 6$)	$\%RSD \leq 2$	0.14	0.11	0.07	0.17

The stability of standard solutions were investigated at intervals of 24 and 48 h. the stability of solutions was determined by comparing area% and peak purity results for PCM, MHB, EHB and PHB. The area% values were within 0.5% after 48 h. These results indicate the solutions were stable for 48 h at ambient temperature, because there was no formation of any unknown or additional peaks. The RSD of peak area, 0.17%; peak purity, 99.98%; asymmetry factor, 1.26 and capacity factor for the peaks were > 2 for each analyte.

3.2 Method Validation

3.2.1 Linearity and range

The test method ability to obtain a response (e.g., peak area) that is directly proportional to the amount of sample tested was evaluated for all components. Linearity was evaluated for PCM, MHB, EHB and PHB in the concentration range 25-160; 20-128; 5-40 and 5-32 µg/mL, respectively, at seven different amounts 25%, 60%, 80%, 100%, 120%, 140 and 160% of analytes ($n = 7$). These solutions were injected in replicates of three. The mean peak areas for replicates ($n = 3$) were plotted against the corresponding injection amounts. The correlation of determination (r^2) obtained were found to be 0.9997, 0.9999, 0.9999 and 0.9998 for PCM, MHB, EHB and PHB, respectively (Table 2). Range is the interval between the upper and lower amount of analyte that can be measured with a specified level of linearity. Range of this method was 25-160% of the routine testing concentration.

Table 2 Validation results for linearity, LOD and LOQ for the HPLC method

Analytes	Conc (μg/mL)	Regression equation	R^2	LOD (μg/mL)	LOQ (μg/mL)
PCM	25-160	y = 34.749x + 266.75	0.9997	(S/N = 3.1) 0.014	(S/N = 10.2) 18
MHB	20-128	y = 38.742x + 595.13	0.9999	(S/N = 3.1) 0.017	(S/N = 10.1) 14
EHB	5-40	y = 18.18x + 29.966	0.9999	(S/N = 3.2) 0.013	(S/N = 10.2) 4
PHB	5-32	y = 12.134x + 45.617	0.9998	(S/N = 3.3) 0.012	(S/N = 10.1) 4

3.2.2 Accuracy/recovery studies

The accuracy of the method was evaluated by adding known quantities of PCM drug and MHB, EHB, PHB preservatives in the suspension formulation samples to give a range of PCM drug and MHB, EHB, PHB preservatives concentration of 80–120% ($n = 3$) of that in a test preparation. These solutions were analysed in triplicate and the amount of analyte recovered calculated. The recovery was 100±2% for all samples with coefficient of variation (CV) less than 2% (Table 3).

TABLE 3 Recovery studies data from sample suspension dosage form

Component	Percent of nominal	n	Taken (mg)	Found (mg)	Recovery (%)
PCM	80, 100, 120	5	120	119.34	99.92-101.07±1.03*
MHB	80, 100, 120	5	80	79.69	99.67-101.04±0.97
EHB	80, 100, 120	5	25	24.87	99.57-101.13±1.14
PHB	80, 100, 120	5	20	19.78	98.96-100.11±1.17

*The coefficient of variation (%CV). %CV as %RSD are both terms describe the same statistical operation.

3.2.3 Specificity

The ability of an analytical method to unequivocally assess the analyte in the presence of other component in the formulation (impurities, degradations, excipients) can be demonstrated by evaluating specificity. The specificity of the method was determined by injecting extracted placebo solutions having the same concentration as that of the sample solution to demonstrate the absence of interference with the elution of the PCM drug and MHB, EHB, PHB preservatives. These results demonstrate (Figure 1b) that there was no interference from the other materials in the suspension formulation and, therefore, confirm the specificity of the method.

Forced degradation studies of the suspension sample were also performed. Suspension samples were prepared and degraded under stress conditions like acidic hydrolysis, basic hydrolysis, oxidative degradation, photo degradation and thermal degradation for LC method. For acid, base and oxidative degradation, samples were individually placed into three volumetric flasks and then 0.1 M HCl, 0.1 M NaOH and 3% H_2O_2 were added separately into the flasks. All the three flasks were then heated in a water bath at 70°C for 4 h acid and base treated sample were neutralised and all the three samples were then diluted to a concentration of 0.1, 0.08, 0.025 and 0.02 mg/mL for PCM, MHB, EHB and PHB, respectively with the mobile phase. For thermal degradation sample was exposed to heat at 105°C for 1 h and for photo degradation, the drug sample was exposed under a UV lamp for 24 h. The samples were withdrawn and analysed using HPLC/UV-DAD.

A summary data of the stress results is shown in Table 4, which showed significant changes in retention times of analytes under acidic, basic and thermal conditions and no degradation peaks were observed under oxidative and photolytic degradation. This was further confirmed by peak purity analysis on a DAD-UV detector (Figure 2) and, therefore, confirms the specificity of the method.

TABLE 4 Summary results of forced degradation study

Stress conditions	Retention Time (min)				Assay (%)			
	PCM	MHB	EHB	PHB	PCM	MHB	EHB	PHB
Reference (fresh solution)	2.60	5.93	8.68	13.77	99.9	99.8	99.9	99.93
Acid degradation	3.28	7.72	11.22	17.85	98.4	99.2	99.2	98.7
Base degradation	3.40	7.78	11.32	18.02	99.1	98.9	99.2	98.8
Oxidative degradation	2.38	5.48	7.98	12.70	99.8	99.8	99.2	99.6
Thermal degradation	2.92	7.12	10.55	17.02	98.7	98.8	98.8	98.8
Photolytic degradation	2.65	5.80	8.25	13.22	99.9	99.9	99.9	99.9

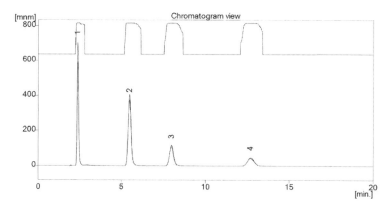

FIGURE 2 HPLC/PDA peak purity chromatogram of PCM (peak 1), MHB (peak 2), EHB (peak 3) and PHB (peak 4).

3.2.4 Precision (intermediate precision and repeatability)

The precision of the method was investigated with respect to repeatability and intermediate precision. Repeatability (intraday precision) of the method was evaluated by assaying six replicate injections of the PCM drug and MHB, EHB, PHB preservatives at 100% of test concentration (0.1, 0.08, 0.025 and 0.02 mg/mL, respectively). The %CV of the retention time (min) and relative percent peak area were found to be < 1%. The intermediate precision (within-laboratory variation) was studied over three consecutive days at three different concentration levels (80, 100, and 120 µg/mL) for PCM, 64, 80, 96 µg/mL for MHB, 20, 25, 30 µg/mL for EHB and 16, 20 and 24 µg/mL for PHB that cover the test method range (80-120%). Three replicate injections were injected for each solution. The grand mean and %CV across the three days and HPLC systems were calculated from the individual peak areas values at the 80%, 100% and 120% testing amounts. The %CV values for peak areas obtained from two days study were < 1% (Figure 3), and met the intermediate precision criteria (CV < 2%) which illustrated the good precision of the analytical method.

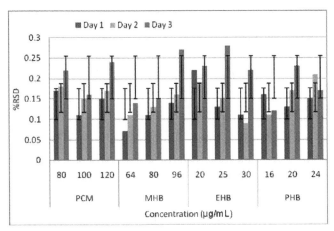

FIGURE 3 Intermediate precision variations evaluated over three days at three different concentration levels. Error bars represent ±1SD.

3.2.5   Limits of detection and quantitation

The limit of detection (LOD) and limit of quantitation (LOQ) tests for the procedure were performed on samples containing very low concentrations of PCM drug and MHB, EHB, PHB preservative. LOD is defined as the lowest amount of analyte that can be detected above baseline noise, typically, three times the noise level. LOQ is defined as the lowest amount of analyte that can be reproducibly quantitated above the baseline noise that gives a signal-to-noise (S/N) ratio of 10. The LOD was (S/N = 3.1) 0.014, (S/N = 3.1) 0.017, (S/N = 3.2) 0.013 and (S/N = 3.3) 0.012 µg/mL and LOQ was (S/N = 10.2) 18, (S/N = 10.1) 14, (S/N = 10.2) 4 and (S/N = 10.1) 4 µg/mL, and %RSD was < 2% (n = 3) for PCM, MHB, EHB and PHB, respectively (Table 2).

4. APPLICATION OF THE METHOD

To demonstrate the applicability of the present method, commercially available three batches of tablets containing 500 mg PCM were analysed. Assay results for three samples of tablets expressed as the percentage of the label claim, were found between 99.59 to 99.97%. Results showed (Table 5) that the content of PCM in tablet formulation was to the counter requirements (90-110% of the label claim). The chromatogram obtained from sample analysis is given in Figure 4. The above results demonstrated that the developed method achieved rapid and accurate determination of compound studied and can be used for the simultaneous determination of PCM drug and MHB, EHB, PHB preservatives in pharmaceutical dosage forms.

TABLE 5 The determination of paracetamol content in tablets dosage form

Component (Lot #)	Label claim (mg)	Found (mg)	n	Recovery (%)
Paracetamol				
1	500	499.87	5	99.97±0.74*
2	500	497.94	5	99.59±0.62
3	500	498.76	5	99.75±1.07

*The coefficient of variation (%CV).

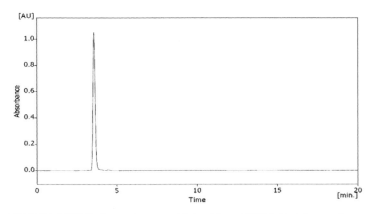

FIGURE 4 HPLC chromatogram obtained from PCM tablet dosage form.

5. CONCLUSION

HPLC/UV-DAD method is simple, rapid, robust and sensitive and therefore suitable for the routine analysis of PCM, MHB, EHB and PHB in suspension and tablet dosage forms as well as raw materials release for production formulation. The proposed method could be used reliably for quality and stability monitoring of PCM, MHB, EHB and PHB.

REFERENCES

1. American Pharmaceutical Association, Handbook of Pharmaceutical Excipients, Washington, D.C. 1986; 17 pp.
2. Shabir, G.A. Determination of combined *p*-hydroxy benzoic acid preservatives in a liquid pharmaceutical formulation and assay by HPLC. *J. Pharm. Biomed. Anal.* 2004, *34*, 207-213.

3. Angelov, T.; Vlasenko, A.; Tashkov, W. HPLC determination of pKa of parabens and investigation on their lipophilicity parameters. *J. Liq. Chromatogr. Relat. Technol.* 2008, *31*, 188-197.

4. Loussouarn, S.; Pouliquen, H.; Armand, F. HPLC determination of oxolinic acid in the plasma of seabass (Dicentrarchus labrax) anesthetized with 2-phenoxyethanol. *J. Chromatogra. B: Biomed. Sci. Appl.* 1997, *698*, 251-259.

5. Gagliardi, L.; De Orsi, D.; Manna, L.; Tonelli, D. Simultaneous determination of antioxidants and preservatives in cosmetics and pharmaceutical preparations by reversed-phase HPLC. *J. Liq. Chromatogr. Relat. Technol.* 1997, *20*, 1797-1808.

6. Sottofatori, E.; Anzaldi, M.; Balbi, A.; Tonello, G. Simultaneous HPLC determination of multiple components in a commercial cosmetic cream. *J. Pharm. Biomed. Anal.* 1998, *18*, 213-217.

7. Shabir, G.A. Method development and validation of preservatives determinations (benzyl alcohol, ethylene glycol monophenyl ether, methyl hydroxybenzoate, ethyl hydroxybenzoate, propyl hydroxybenzoate and butyl hydroxybenzoate) using HPLC. *J. Liq. Chromatogr. Relat. Technol.* 2007, *30*, 1951-1962.

8. Serrano, F.O.; Lopez, I.S.; Revilla, G.N. HPLC determination of chemical preservatives in Yogurt. *J. Liq. Chromatogr. Relat. Technol.* 1991, *14*, 709-717.

9. Khan, S.H.; Murawski, M.P.; Sherma, J. Quantitative high performance thin layer chromatographic determination of organic acid preservatives in beverages, *J. Liq. Chromatogr. Relat. Technol.* 1994, *17*, 855-865.

10. Shabir, G.A.; Lough, W.J.; Shafique, A.A.; Shar, G.Q. Method development and validation for the HPLC assay of phenylformic acid, 2,4-hexadienoic acid, methyl 4-hydroxybenzoate and propyl 4-hydroxybenzoate. *J. Liq. Chromatogr. Relat. Technol.* 2006, *29*, 1223-1233.

11. Mannucci, C.; Bertini, J.; Cocchini, A.; Perico, A.; Salvagnini, F., Triolo, A. HPLC simultaneous quantitation of ketoprofen and parabens in a

commercial gel formulation. *J. Liq. Chromatogr. Relat. Technol.* 1992, *15*, 327-335.

12. Padmanabhan, G.R.; Smith, J.; Mellish, N.; Fogel, G. HPLC determination of parabens in pharmaceutical preparations containing hydroxyquinolines. *J. Liq. Chromatogr. Relat. Technol.* 1982, *5*, 1357-1366.

13. Mahuzier, P.E.; Altria, K.D.; Clark, B.J. Selective and quantitative analysis of 4-hydroxybenzoate preservatives by microemulsion electrokinetic chromatography. *J. Chromatogr. A*, 2001, *924*, 465-470.

14. Lee, M.R.; Lin, C.Y.; Li, Z.G.; Tsai, T.F. Simultaneous analysis of antioxidants and preservatives in cosmetics by supercritical fluid extraction combined with liquid chromatography-mass spectrometry. *J. Chromatogr. A*, 2006, *1120*, 244-251.

15. Grosa, G.; Grosso, E.D. Russo, R.; Allegrone, G. Simultaneous, stability indicating, HPLC-DAD determination of guaifenesin and methyl and propyl-parabens in cough syrup. *J. Pharm. Biomed. Anal.* 2006, *41*, 798-803.

16. Altun, M.L. HPLC method for the analysis of paracetamol, caffeine and dipyrone. *Turk. J. Chem.* 2002, *26*, 521-528.

17. Cenyuva, H.; Qzden, T. Simultaneous HPLC determination of paracetamol, phenylephrine HCl, and chlorpheniramine maleate in pharmaceutical dosage forms. *J. Chromatogr. Sci.* 2002, *40*, 97-100.

18. Shabir, G.A.; Lough, W.J.; Shafique, A.A.; Bradshaw, T.K. Evaluation and application of best practice in analytical method validation. *J. Liq. Chromatogr. Rel. Technol.* 2007, *30*, 311-333.

19. Shabir, G.A. A practical approach to validation of HPLC methods under current good manufacturing practices. *J. Validation Technol.* 2004, *10*, 210-218.

20. Shabir, G.A. Validation of HPLC methods for pharmaceutical analysis: Understanding the differences and similarities between validation requirements of the U.S. Food and Drug Administration, the U.S. Pharmacopoeia and the International Conference on Harmonization. *J. Chromatogr. A*, 2003, *987*, 57-66.

21. Shabir, G.A. Step-by-step analytical methods and protocol in the quality system compliance industry. *J. Validation Technol.* 2004, *10*, 314-324.

22. ICH, Q2 (R1), International Conference on Harmonization, Validation of analytical procedures: Text and methodology, Geneva, Switzerland, 2005.

23. U.S. Pharmacopeia 34. <621>, Chromatography, United States Pharmacopieal Convention: Rockville, Maryland, 2010.

24. Shabir, G.A. Systematic strategies in high-performance liquid chromatography method development and validation. *Sep. Sci. Technol.* 2010, *45*, 670-680.

13 DEVELOPMENT AND VALIDATION OF A HPLC METHOD FOR THE DETERMINATION OF GUAIPHENESIN AND SODIUM BENZOATE IN LIQUFRUTA GARLIC COUGH MEDICINE

ABSTRACT

A new and simple isocratic high-performance liquid chromatographic method with ultraviolet detection is described for simultaneous determination of active guaiphenesin and preservative sodium benzoate in liqufruta garlic cough medicine formulation. The chromatographic separation was achieved using a Zorbax CN; 150 mm × 4.6 mm and 5 μm particle size column employing acetonitrile and water (20:80, v/v) containing 0.1% formic acid (pH 3.5 ±0.05) as the mobile phase. The method was validated with respect to linearity, range, precision, accuracy, specificity, limit of detection and limit of quantitation. The both analytes were detected by UV-Vis detector at 245 nm. The method was liner over the concentration range of 0.2 – 0.8 mg/mL and 0.02 – 0.06 mg/mL for guaiphenesin and sodium benzoate, respectively. The limit of detection was found to be 0.14 μg/mL for GP and 0.06 μg/mL for SB and the quantification limit was 0.54 μg/mL for GP and 0.22 for SB. Accuracy, evaluated as recovery, was in the range of 97.8 – 100.02%. Intra-day precision and intermediate precision showed relative standard deviation < 1% in each case.

Keywords: HPLC, Method development, Method validation, Guaiphenesin, Sodium benzoate

1. INTRODUCTION

Guaiphenesin (GP, Figure 1a) also known as 3-(2-methoxyphenoxy)-propane-1,2-diol is widely used in many medicinal liquid formulations as an active ingredient. It is available alone or in combination with other drugs, mainly as syrup in cough-cold products. GP is used to treat symptoms of allergy, colds and upper respiratory infections. Sodium benzoate (SB, Figure 1b) is commonly used single preservative, but more often combinations of preservatives as anti-fungal and antimicrobial agents in cosmetics, beverages, food and pharmaceutical formulations to prevent alteration and degradation of the product formulations.

Many existing analytical procedures are available in literature for the determination of

GP either individually or in combination with other drugs by high-performance liquid chromatography (HPLC) and other analytical techniques [1-9]. The novelty of the proposed method is the simultaneous analysis of GP drug and a combination of the SB preservative in liqufruta garlic cough medicine formulation.

The aim of this study was to develop and critically validate a new simple HPLC method with ultraviolet detection for the simultaneous determination of GP and SB compounds in a single chromatographic run from medicinal dosage forms. Analytical methods validation is an important regulatory requirement in pharmaceutical analysis. In recent years, the International

242

Conferences on Harmonization (ICH) has introduced Guidelines for analytical methods validation [10]. The most widely applied analytical performance characteristics are accuracy, specificity, linearity, range, precision (repeatability and intermediate precision), limit of detection and limit of quantitation. The purpose of this study was to develop and validate a rapid, accurate, simple and robust reversed-phase HPLC method for the quantitation of GP in combination with SB preservative in a single chromatographic run which can be reliably used in routine quality control analysis for raw materials, bulk and final products release.

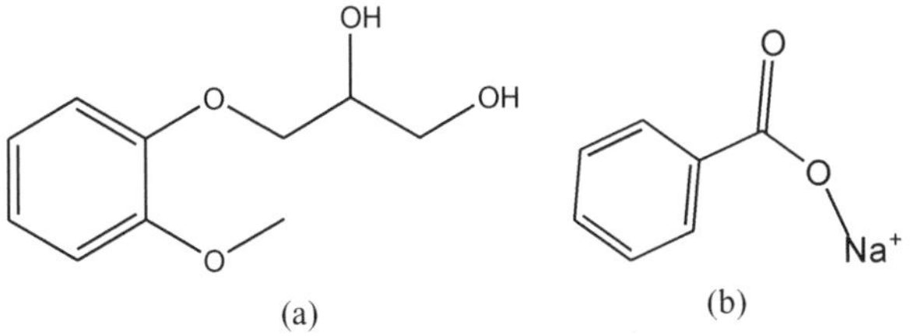

FIGURE 1 Chemical structures of (a) guaiphenesin (GP); and (b) sodium benzoate (SB).

2. EXPERIMENTAL

2.1 Materials

Acetonitrile (HPLC-grade), formic acid (analytical grade), guaiphenesin (pure 99%) and sodium benzoate (pure 99.5%) were purchased from Sigma-Aldrich (Gillingham, UK). Distilled water was de-ionised by using a Milli-Q system (Millipore, Bedford, MA).

2.2 HPLC Instrumentation and Conditions

The chromatographic separation was carried out on a Knauer HPLC system (Berlin, Germany) equipped with a model 1000 LC pump, model 3950 autosampler, model 2600 photodiode-array (PDA) detector and a vacuum degasser was used. The data were acquired via Knauer ClarityChrom Workstation data acquisition software. Reversed-phase HPLC analysis was performed isocratically at 25°C using a Zorbax CN (150 mm × 4.6 mm, 5 μm) column (Hichrom, Reading, UK). The mobile phase consisted of a mixture of acetonitrile and water (20:80, v/v) containing 0.1% formic acid adjusted to pH 3.5 ±0.05 with 2 M sodium hydroxide was used. The flow rate was 1.0 mL/min and injection volume was 10 μL. The eluent was monitored with a UV detector set at 245 nm. All samples were diluted with mobile phase.

2.3 Standard and Sample Preparation

Accurately weighted (1.0 g) GP in 100 mL volumetric flask and diluted in methanol (S1). Accurately weighted (0.162 g) SB in 250 mL volumetric flask and dissolved in water (S2). 5 mL aliquot of S1 and 5 mL aliquot of S2 were added to the same 100 mL volumetric flask, and diluted in mobile phase, yielding a final concentration of 0.50 mg/mL and 0.0324 mg/mL, respectively. This solution was filtered through 0.45 micron filter before injecting into the HPLC system.

Accurately weighed (6.5 g) sample in 100 mL volumetric flask and diluted in mobile phase. This sample solution was filtered through 0.45 micron nylon filter before injecting into the HPLC system.

Linearity experiments were performed by preparing GP and SB standard solutions in the range 200-800 µg/mL and 20-60 µg/mL in mobile phase, respectively and injected in triplicate. Linear regression analysis was carried out on the standard curve generated by plotting the concentration of GP and SB versus peak area response. Placebo formulation was prepared mixing all excepients without adding active drug GP and preservative SB. This placebo sample was prepared and injected in duplicate to see any interference with the active and preservative peaks from any exceipents.

3. RESULTS AND DISCUSSION

3.1 Method Development

The chromatographic separation of GP and SB was carried out in isocratic mode using a mixture of acetonitrile-water pH 3.5 ±0.5 (20:80, v/v) containing 0.1% formic acid as mobile phase. The column was equilibrated with the mobile phase flowing at 1.0 mL/min for about 30 min prior to injection. Column temperature was held at 25°C although separation at 20°C and 30°C indicated that slight variation in temperature did not have a significant effect on retention or peak shape. Therefore, results produced at 25°C were more reproducible and stable comparing 20°C and 30°C, so 25°C was selected as the working temperature. The injection volume of 10 μL of standard solutions was injected automatically into the column. Subsequently, the liquid chromatographic behavior of the preservative was monitored with a PDA UV detector at 245 nm. Additionally, preliminary system suitability, precision, linearity, robustness and stability of the solution studies performed during the development phase of the method showed that the 10 μL injection volumes was reproducible and the peak response was significant at the analytical concentration chosen. In the optimized conditions, the drug substance and preservative were well separated with a resolution of more than 7.6 (Figure 2).

System suitability testing was performed to determine the accuracy and precision of the system from six replicate injections of solutions containing 0.50 mg GP/mL and 0.0324 mg SB/mL. The percent relative standard deviation (%RSD) of the retention time (min) and peak area were found to be less than 0.14%. The retention factor (also called capacity factor, k) was calculated using the equation $k = (t_r / t_0) -1$, where t_r is the retention time of the analyte and t_0 is the retention time of an unretained compound; in this study, t_0 was calculated from the first disturbance of the baseline after injection and capacity factor value was obtained 3.23 and 6.80 for GP and SB peaks. The separation factor (α) was calculated using the equation, $\alpha = k_2 / k_1$ where k_1 and k_2 are the retention factors for the first and last eluted peaks respectively. The separation factor for GP and SB peaks 2.56 and 5.54 obtained. The plate number (also known as column efficiency, N) was calculated as $N = 5.54 \, (t_r / w_{0.5})^2$ where $w_{0.5}$ is the peak width at half peak height. In this study, the theoretical plate number for GP and SB were 4839 and 6033 respectively. Resolution is calculated from the equation $R_s = 2(t_2 - t_1) / (t_{w1} + t_{w2})$. Where t_1 and t_2 are retention times of the first and second eluted peaks, respectively, and t_{w1} and t_{w2} are the peak widths. The resolution between SB and GP peaks was > 7.60. The asymmetry factor (A_s) was calculated using the US Pharmacopeia (USP) method. The peak asymmetry value for GP and SB peaks were 1.25 and 1.20 respectively. The proposed method met these requirements within the accepted limits [11-15].

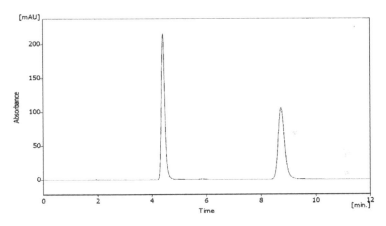

FIGURE 2 HPLC chromatogram of active drug GP and preservative SB. (1) GP, RT 4.42 min, sample concentration 0.50 mg/mL, (2) SB, RT 8.73 min, sample concentration 0.0324 mg/mL.

Robustness studies were also performed in the method development phase applying the experimental design as shown in Table 1. A sample of GP and SB were prepared at the working concentration (0.50 and 0.0324 mg/mL) and assayed using the experimental design with eight test combinations for seven different chromatographic parameters as shown in Table 2. For each parameter, four combinations of (AAAA) and four combinations of (aaaa) were studied. The actual value of each parameter (VA–VG) (Table 2) shows which parameter has a dominant influence on the developed analytical method. In all cases, good separations of GP and SB were always achieved, indicating that the analytical method remained selective for the GP and SB components under the optimized conditions.

TABLE 1 Experimental design for robustness study

Test parameter	1	2	3	4	5	6	7	8
A/a	A	A	A	A	a	a	a	a
B/b	B	B	b	b	B	B	b	b
C/c	C	c	c	c	C	c	C	c
D/d	D	D	d	d	d	d	D	D
E/e	E	e	E	e	e	E	e	E
F/f	F	f	f	F	F	f	f	F
G/g	G	g	g	G	g	G	G	g
Results	s	t	u	v	w	x	y	z

TABLE 2 Chromatographic parameter for robustness study

Parameter	Test conditions 1	Test conditions 2	Differences
Analytical column	A = column Zorbax CN	a = column Zorbax CN	$V_A = \frac{1}{4}(s+t+u+v) - \frac{1}{4}(w+x+y+z) = A - a$
Sample solvent	B = Buffer/water	b = Mobile phase	$V_B = \frac{1}{4}(s+t+w+x) - \frac{1}{4}(u+v+y+z) = B - b$
Temperature	C = 20°C	c = 30°C	$V_C = \frac{1}{4}(s+u+w+y) - \frac{1}{4}(t+v+x+z) = C - c$
Flow rate	D = 0.8 mL/min	d = 1.2 mL/min	$V_D = \frac{1}{4}(s+t+y+z) - \frac{1}{4}(u+v+w+x) = D - d$
Wavelength	E = 235 nm	e = 255 nm	$V_E = \frac{1}{4}(s+u+x+z) - \frac{1}{4}(t+v+w+y) = E - e$
Mobile phase	F = 18% acetonitrile	f = 22 % acetonitrile	$V_F = \frac{1}{4}(s+v+w+z) - \frac{1}{4}(t+u+x+y) = F - f$
Solubility stability	G = 1 h	g = 48 h	$V_G = \frac{1}{4}(s+v+x+y) - \frac{1}{4}(t+u+w+z) = G - g$

3.2 Validation of the HPLC Method

3.2.1 Linearity and range

The linearity test was performed using five different amounts of GP and SB analytes in the range 200-800 µg/mL and 20-60 µg/mL, respectively covering the target concentration (0.50 GP and 0.0324 SB mg/mL). Solutions corresponding to each concentration level were injected in triplicate. This design is consistent with the ICH Guideline [10], which recommends a minimum of five injection levels to establish linearity. The results are presented in Table 3 and show a good correlation between the peak area of analytes and concentration with r > 0.9998.

TABLE 3 Validation linearity assessment results for HPLC method

Analytes	Conc	Range	Equation for regression line	r^2
GP	500 (µg/mL)[a]	200-800 (µg/mL)	y = 8000x - 1167.6	0.9998
SB	32.4 (µg/mL)[a]	20-60 (µg/mL)	y = 62380x - 729.6	0.9999

[a] Target concentration corresponding to 100%

3.2.2 Precision studies

The precision of the method was determined by repeatability (intra-day) and intermediate precision (inter-day variation). Repeatability was examined by analysing six determinations of the same batch of each component at 100% of the test concentration (0.50 GP and 0.0324 SB mg/mL). The %RSD of the areas of preservative peak were found to be less than 0.22% (Table 4), which confirms that the method is sufficiently precise. Intermediate precision (inter-day variation) was studied by assaying five samples containing the nominal amount of GP and SB on different days by different analysts using different LC systems. Solutions corresponding to each concentration level were injected in duplicate. The %RSD values across the systems were calculated and found to be less than 0.52% (Table 4) for each of the multiple sample preparations, which demonstrates excellent precision for the method.

TABLE 4 Validation results for HPLC method

Validation steps	Parameters	Results		Acceptance criteria
		GP	SB	
Repeatability	RSD (%, $n = 6$)	0.14	0.19	X < 2
Intermediate precision				
Day 1, LC 1, analyst 1	RSD (%, $n = 5$)	0.24	0.23	X < 2
Day 2, LC 2, analyst 2	RSD (%, $n = 5$)	0.17	0.30	X < 2
Stability (24 h)	Change in response factor (%)	0.09	0.22	X < 2

3.2.3 Accuracy/recovery studies

Recovery studies may be performed in a variety of ways depending on the composition and properties of the sample matrix. In the present study, three different solutions were prepared with a known amount of pure GP and SB added to give a concentration range of 80%, 100% and 120% of that in a test preparation covering the assay range. These solutions were injected in triplicate and percent recoveries of response factor (area/concentration) were calculated (Table 5).

TABLE 5 Recovery studies data for HPLC method

Components	Applied concentration (% of target)		
	80%	100%	120%
GP	100.01 (± 0.11)[a]	99.98 (± 0.18)	99.89 (± 0.17)
SB	99.98 (± 0.20)	100.02 (± 0.09)	97.82 (± 0.26)

[a] The coefficient of variation, (n = 3).

3.2.4 Specificity

The LC-DAD purity isoplot chromatogram (Figure 3) demonstrates a good separation of the GP (RT = 4.42 min), and SB (RT = 8.73 min) from each other. A wavelength of 245 nm was found to be the most effective compromise to accomplish the detection and quantification of the two components in a single run. The GP and SB peaks were adequately resolved from each other, typical resolution values were > 7.60. Therefore, this method demonstrates acceptable specificity. Injections of the extracted placebo were performed to demonstrate the absence of interference with the elution of the GP and SB analytes. This result demonstrates (Figure 4) that there was no interference from the other materials in the liqufruta garlic cough formulation and, therefore, confirms the specificity of the method.

Forced degradation studies were performed to evaluate the specificity of each compound under four stress conditions (heat, UV light, acid, base). Solutions of each component were exposed to 60°C for 1 h, UV light using a UVL-56 lamp for 24 h, acid (1M hydrochloric acid) for 24 h, and base (1M sodium hydroxide) for 4 h.

A summary data of the stress results is shown in Table 6, which showed no changes in retention times of each compound by peak purity analysis on a DAD UV detector and, therefore, confirms the specificity of the method.

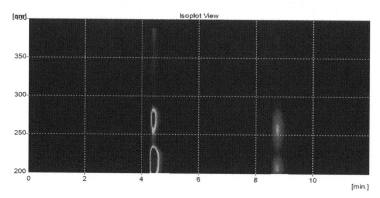

FIGURE 3 LC-DAD/UV isoplot chromatogram of GP and SB.

TABLE 6 Summary of forced degradation studies for HPLC method

Stress conditions	Sample treatment	t_R (min)		Assay (%)	
		GP	SB	GP	SB
Reference	Fresh solution	4.42	8.73	99.98	99.97
Acid degradation	1M HCl for 24 h	4.43	8.74	99.92	99.95
Base degradation	1M NaOH for 4 h	4.43	8.74	99.94	99.88
Heat degradation	60°C for 1 h	4.42	8.73	100.00	99.97
Light degradation	UV Light for 24 h	4.42	8.73	99.93	99.96

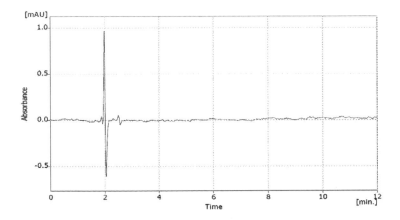

FIGURE 4 HPLC chromatogram obtained from placebo formulation, demonstrates the absence of interference with the elution of the GP and SB. (See standard and sample preparation section for solution used).

3.2.5   Limit of detection and limit of quantitation

The limit of detection (LOD) is the lowest amount of analyte in a sample that can be detected, but not necessary quantitated as an exact value. To determine the minimum level at which the analyte can be reliably detected was established by measuring the signal to noise (S/N) ratio. A signal to noise ratio of 3:1 is acceptable [10].

The limit of quantitation (LOQ) is the lowest amount of analyte in a sample that can be determined with acceptable precession and accuracy under the stated experimental conditions. The LOQ is expressed as the concentration of analyte in the sample. To determine the minimum level at which the analyte can be reliably quantitated was established by measuring the signal to noise (S/N) ratio. A signal to noise ratio of 10:1 is acceptable [10].

The LOD and LOQ of GP and SB were determined based on standard deviation (σ) of response and slope(s). GP and SB solutions were prepared in the range 0.04 -200 µg/mL and 0.008 - 20 µg/mL and injected in triplicate. Average peak areas of analytes were plotted against concentration. LOD and LOQ were calculated by using the following equations: LOD = (3.3 σ)/s; and LOQ = (10 σ)/s. The LOD was determined to be 0.14 µg/mL (s/n = 3.2) for GP and 0.06 µg/mL (s/n = 3.5) for SB and LOQ was found to be 0.54 µg/mL (s/n = 10.6) for GP and 0.22 µg/mL (s/n = 11.04) for SB with %RSD < 2% for six replicate injections for both analytes.

3.2.6 Stability of analytical solutions

Sample solutions were chromatographed immediately after preparation and then re-assayed after storage at room temperature for 24 h. The results given in Table 4 showed there was no significant change (< 0.13% response factor) in GP and SB concentrations (0.50 and 0.0324 mg/mL) over this period.

4. CONCLUSION

A simple and rapid reversed-phase HPLC method with UV spectrophotometric detection was developed for the simultaneous determination of active GP and preservative SB in liqufruta garlic cough medicine formulation. The retention times of the drugs were found to be 4.42 min and 8.73 min for GP and SB, respectively. The method was validated and the results obtained were accurate and precise with RSD < 1% in all cases and no significant interfering peaks were detected. The method is specific, selective, robust and reliable for routine use in quality control analysis of GP and SB raw materials and final product release for marketing authorisation.

REFERENCES

1. Pappano, N.B., Micalizzi, Y.C., Debattista, N.B., and Ferretti, F.H., *Talanta,* 1997, vol. 44, no. 4, p. 633.
2. Lee, A.R., and Hu, T.M., *J. Pharm. Biomed. Anal.,* 1994, vol. 12, no.6, p. 747.
3. Abdel-Hay, M.H., El-Din, M.S., and Abuirjeie, M.A., *Analyst,* 1992, vol. No.2, 117, p. 157.
4. Tapsoba, I., Belgaied, J.E., and Boujel, K., *J. Pharm. Biomed. Anal.,* 2005, vol. 38, no. 1, p. 162.
5. Pomponio, R., Gotti, R., Hudaib, M., Andrisano, V., and Cavrini, V., *J. Sep. Sci.,* 2001, vol. 24, no. 2, p. 258.
6. Grosa, G., Del Grosso, E., Russo, R., and Allegrone, G., *J. Pharm. Biomed. Anal.,* 2006, vol. 41, no. 3, p. 798.
7. Indrayanto, G., Sunarto, A., and Adrian, Y., *J. Pharm. Biomed. Anal.,* 1995, vol. 13, no. 12, p. 1555.
8. Wilcox, M.L., and Stewart, J.T., *J. Pharm. Biomed. Anal.,* 2000, vol. 23, no. 5, p. 909.
9. Vasudevan, M., Ravisankar, S., Sathiyanarayanan, A., and Chandan, R. S., *J. Pharm. Biomed. Anal.,* 2000, vol. 24, p. 25.
10. International Conference on Harmonization (ICH), Validation of Analytical Procedures, Q2 (R1), 2005.

11. Reviewer Guidance: Validation of Chromatographic Methods, Food and Drug Administration (FDA), Center for Drug Evaluation and Research (CDER), 1994.

12. Shabir, G,A., *J. Chromatogr. A,* 2003 vol. 987, no. 1-2, p. 57.

13. Shabir, G.A., Lough, W.J., Shafique, A.A., and Bradshaw, T.K., *J. Liq. Chromatogr. Relat. Technol.,* 2007, vol. 30, no. 3, p. 311.

14. Shabir, G.A., *J. Validation Technol.,* 2004, vol. 10, no. 4, p. 314.

15. Shabir, G.A., *Pharma. Technol. Europe*, 2004, vol. 16, no. 3, p. 37.

14 DEVELOPMENT AND VALIDATION OF STABILITY-INDICATING HPLC METHOD FOR THE DETERMINATION OF DOMPERIDONE, SORBIC ACID AND PROPYLPARABEN IN PHARMACEUTICAL FORMULATION

ABSTRACT

A simple, selective and sensitive stability-indicating LC method has been developed and validated for the simultaneous determination of sorbic acid, propylparaben and domperidone in pharmaceutical oral suspension formulations. The separations was achieved on a Lichrosorb C_8, 150 mm × 4.6 mm and 5 μm column with detection of 280 nm using an isocratic mobile phase mixture of phosphate buffer (0.05 M) and methanol (40:60 v/v) at flow rate of 1.0 mL/min. Under these conditions, separation of the three components was achieved in less than 10 min. The retention times for sordid acid, propylparaben and domperidone were found to be 3.88, 6.12 and 8.53 min with good resolution of 7.90 and 6.11, respectively. The calibration curve for sorbic acid, propylparaben and domperidone was linear in the range of 30-150, 5-30 and 36-180 μg/mL, respectively with $r = \geq 0.9998$ for each component. The proposed method was successfully employed for quantification of sorbic acid, propylparaben and domperidone in pharmaceutical formulations.

Keywords: LC, Stability-indicating, Domperidone, Sorbic acid, Propylparaben, Pharmaceutical preparation

1. INTRODUCTION

Domperidone (DP) (4-(5-chloro-2-oxo-1-benzimidazolinyl)-1-[3-(2-oxobenzimidazolinyl) propyl]piperidine, $C_{22}H_{24}ClN_5O_2$, Figure 1) is an antidopaminergic drug used in tablet, oral suspension and suppositories formulations. It stimulates gastro-intestinal motility and is used as an antiemetic for the short term treatment of nausea and vomiting of various aetiologies, including that associated with cancer therapy and with levodopa or bromocriptine therapy for parkinsonism [1] Some organic acids and their esters are commonly used single preservatives, but more often combinations of preservatives as antimicrobial agents in cosmetic, food and pharmaceutical products [2] to prevent chemical alteration and degradation of the product formulation. Sorbic acid (SA) is generally effective to control mold and inhibit yeast growth and against a wide range of bacterial attack [3] Propylparaben (PP) is the most commonly used preservative and have been used for many years. It had been found that the antimicrobial activities of the parabens seem to increase with increasing chain length. However, esters of longer alkyl chains are of limited applications due to their lower solubility in water [4]. The analysis of these preservatives in commercial pharmaceutical products is particularly important both for quality assurance and consumer protection. Domperidone is also known as a dopamine D_2 receptor antagonist used as antiemetic agent into human beings for preventing nausea and vomiting. It is being used widely all over the world for its unique pharmaceutical activity. Therefore, the analysis of DP into oral suspensions in combination with preservatives is required and urgent needed. Some LC methods are available on the determination of DP, either alone or in combination with other drugs [5-8]. Several analytical procedures have been reported for the determination of SA and PP preservatives separately or in combination with other drugs by LC and other techniques [9-21]. These methods may not be suitable for simultaneous determination of SA, PP and DP together in one chromatographic run. However, as per literature search performed, no LC method has been found in oral suspension for simultaneous determination of these compounds containing combination of three components together SA, PP and DP. Therefore, attempts are made to develop fast, sensitive, selective and robust method for simultaneous determination of SA, PP and DP into in oral suspension formulations. The present research describes the analysis of SA, PP and DP, which is in the pharmaceutical oral suspension formulations by liquid chromatography.

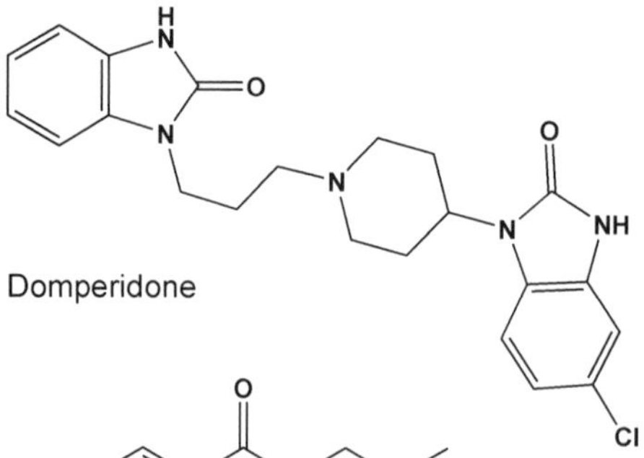

Domperidone

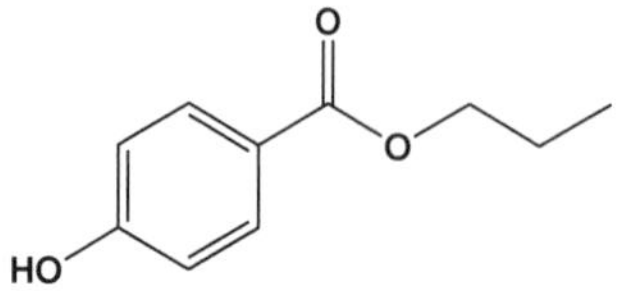

Propylparaben (Propyl 4-hydroxybenzoate)

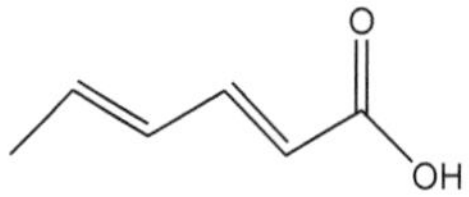

Sorbic acid (2,4-hexadienoic acid)

FIGURE 1 Chemical structures of the separated compounds.

2. EXPERIMENTAL

2.1 Chemicals and Reagents

Methanol (HPLC-grade), domperidone ($\geq$ 98% HPLC), sorbic acid ($\geq$ 99.0%), propylparaben ($\geq$ 99.0 %) and potassium dihydrogen phosphate ($\geq$ 99.0%) were purchased from Sigma-Aldrich (Gillingham, UK). Ortho-phosphoric acid (85%) was obtained from Merck Chemicals (Nottingham, UK). Purified water was prepared by Milli-Q system (Bedford, MA, USA).

2.2 Instrumentation and Vonditions

Chromatographic separation was carried out on a Knauer HPLC system (Berlin, Germany) equipped with a model 1000 LC pump, model 3950 autosampler, model 2600 photodiode-array (PDA) detector and a vacuum degasser was used. The data were acquired via Knauer ClarityChrom Workstation data acquisition software. The mobile phase consisted of a mixture of phosphate buffer (0.05 M) and methanol (40:60, v/v) was used. The flow rate was set to 1.0 mL/min. The injection volume was 20 µL and the detection wavelength was set at 280 nm. Reversed-phase LC analysis was performed isocratically at 30°C using a Lichrosorb C_8 (150 × 4.6 mm, 5 µm) column (Jones Chromatography, Hengoed, UK).

2.3 Standard Preparation

An accurately weighed amount (20 mg) of PP was placed in a 100 mL volumetric flask and dissolved in methanol to produce a standard solution (S1). An accurately weighed amount (12 mg) of DP and (10 mg) of SA were transferred into 100 mL volumetric flask. A 50 mL of mobile phase was added and dissolved. A 10 mL aliquot of stock solution S1 was added, and volume was competed with mobile phase, yielding a final concentration of 0.12 mg of DP, 0.10 mg of SA and 0.02 mg of PP/mL.

2.4 Sample Preparation

An accurately weighed amount (10.0 g) of sample suspension was transferred into 100 mL volumetric flask. A 50 mL of mobile phase was added. This mixture was subject to sonication for 10 min for complete extraction of drug and preservatives and the solution was made up to the mark with mobile phase. The solution was centrifuged at 4000 rpm for 5 min; the clear supernatant portion was collected and filtered through a 0.22 μm membrane filter (Millipore, Watford, UK) and 20 μL of this solution was injected onto the HPLC system.

3. RESULTS AND DISCUSSION

3.1 Optimization of Chromatographic Conditions

For chromatographic separation of SA, PP and DP a Lichrosorb C_8 stationary phase (4.6 mm, 5 μm) with varying column lengths from 250 to 150 mm were attempted. Different mobile phase compositions containing phosphate buffer (0.05 M) and methanol (50:50, 40:60, v/v) were tried. Although good separation was achieved with phosphate buffer and methanol in the ratio of 40:60 (v/v), DP peak symmetry was found to be greater than 2.0. The symmetry of the DP peak was improved by addition of 0.05% phosphoric acid in the mobile phase. The chromatographic separation with better peak shape was achieved using a mixture of aqueous 0.05% phosphoric acid and methanol in the ratio of 40:60 (v/v). The column, 250 mm × 4.6 mm, 5μm showed higher elution time (4.46 for SA, 6.72 for PP and 9.12 min for DP) with resolutions of 8.15 for PP and 7.73 for DP. The shorter column length (150 mm × 4.6 mm, 5µm) has reduced the elution time (3.88 for SA, 6.12 for PP and 8.53 min for DP) with good resolution (7.90 and 6.11 USP), respectively. The overlaid photodiode-array (PDA) spectrum showed good response at 280 nm for all three components. Therefore, this wavelength was used for simultaneous determination of drug and both preservatives. In the optimized conditions, SA, PP and DP were separated with a resolution of > 6 and the retention times were found to be 3.88, 6.12 and 8.53 min, respectively. Chromatogram of system suitability and suspension sample are shown in Figure 2 and 3.

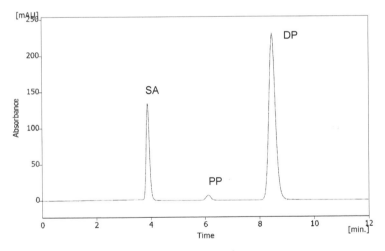

FIGURE 2 LC chromatogram obtained from standard during system suitability experiments.

The system suitability test was performed to confirm that the LC system to be used was suitable for the intended application. A standard solution containing of 0.10 mg of SA, 0.02 mg of PP and 0.12 mg of DP/mL was injected six times. The parameters measured were peak area, retention time, capacity factor, theoretical plate and tailing factor. The RSD values calculated for the peak area were 0.43, 0.52 and 0.37 and retention times were 0.09, 0.07 and 0.06 for SA, PP and DP, respectively. The tailing factors were 1.28, 1.40 and 1.15 for SA, PP and DP, respectively. Theoretical plates were 4798, 5262, 5756 and the resolutions were 4.96, 7.91 and 6.11 for SA, PP and DP, respectively. The system suitability experimental results showed that the parameters evaluated were within the acceptable range (RSD < 2.0%) indicating that the system was suitable for the analysis intended applications.

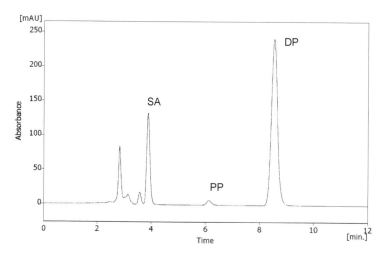

FIGURE 3 LC chromatogram obtained from pharmaceutical oral suspension sample.

The robustness of an analytical procedure refers to its ability to remain unaffected by small and deliberate variations in method parameter and provides an indication of its reliability for the routine analysis. To determine the robustness of the method the experimental conditions were deliberately altered and retention times (min), assay percent, peak tailing, number of theoretical plates and resolutions were evaluated.

The mobile phase flow rate was 1.0 mL/min. this was changed by 0.1 units to 0.9 and 1.1 mL/min and the effect was studied. Similarly, the effect of column temperature was studied at 28°C and 32°C instead of 30°C. The effect of mobile phase composition was studied by use of phosphate buffer (0.05 M) and methanol 38:62 and 42:58 (v/v). The effect of detection wavelength was studied at 275 and 285 nm. For all changes in conditions the sample was analyzed in triplicate. When the effect of altering one set of conditions was tested, the other conditions were held constant at the optimum values. Assay of SA, PP and DP for all deliberate changes of conditions was within 98.09-99.90%. The summary of results is shown in Table 1.

The solution stability of SA, PP and DP in the assay method was investigated by leaving sample test solutions in tightly capped volumetric flasks at room temperature for 48 h. the same sample solutions were assayed at 6 h intervals up to the end of the study period against freshly prepared standard solutions. The RSD (%) of the assay of SA, PP and DP were calculated for the study period during solution stability experiments. The RSD values of the assay of SA, PP and DP were less than 1.0%. No significant changes were observed during solution stability experiments. The results from these experiments confirm that sample solutions used during assay were stable up to the study period of 48 h.

TABLE 1 Result from evaluation of the robustness study of the LC method

Condition	Retention time (min)	Assay (%)	USP peak tailing	USP resolution	Theoretical plates
Flow rate (±10% of the optimum flow)					
0.9 mL/min	3.9, 6.2, 8.6	99.6, 99.9, 99.8	1.1, 0.7, 1.0	7.9, 6.2	4699, 5182, 5673
1.1 mL/min	3.7, 6.0, 8.4	99.7, 99.8, 99.5	1.0, 0.8, 0.9	7.8, 6.0	4684, 5168, 5665
Mobile phase composition (±2% of optimum organic modifier concentration)					
58 mL	4.0, 6.4, 8.8	99.8, 99.2, 99.7	1.0, 0.8, 1.0	7.8, 6.3	4694, 5172, 5678
62 mL	3.3, 5.9, 8.2	99.5, 98.8, 99.9	1.1, 0.9, 1.0	7.9, 6.2	4714, 5245, 5732
Temperature (±2°C of optimum temperature)					
28°C	3.8, 6.1, 8.4	99.7, 99.3, 99.8	1.0, 0.8, 1.0	7.8, 6.5	4689, 5179, 5677
32°C	3.9, 6.4, 8.7	99.4, 99.0, 99.5	0.9, 0.7, 0.9	7.7, 6.2	4694, 5176, 5682
Wavelength (±5 nm of the optimum wavelength)					
275 nm	3.9, 6.3, 8.6	99.2, 99.6, 99.2	1.2, 0.9, 1.0	7.6, 6.3	4699, 5180, 5675
285 nm	3.7, 6.1, 8.4	99.8, 99.2, 99.9	1.0, 0.8, 1.0	7.8, 6.2	4689, 5178, 5652

n = 3 determinations, data for SA, PP and DP, respectively

3.2 Method Validation

The proposed method was validated with respect to linearity and range, specificity, precision, accuracy, limit of detection, limit of quantitation, robustness and stability of analytical solutions following International Conference on Harmonization (ICH) [22], Shabir G.A [23] and United States Pharmacopeia (USP) [24] guidelines.

3.2.1 Linearity and range

Linearity test solutions were prepared by diluting stock solutions of SA, PP and DP (1 mg/mL) at six concentration levels from 30 to 150% levels of analytes concentration (30-150, 5-30 and 36-180 µg/mL, respectively). The solutions were injected in triplicate and the following regression equations were found by plotting the peak area (y) versus the SA, PP and DP concentration (x) expressed in mg/mL:

$y_{SA} = 11.576x + 69.486$ ($R^2 = 0.9999$)

$y_{PP} = 12.721x - 14.737$ ($R^2 = 0.9998$)

$y_{DP} = 31.667x - 403.5$ ($R^2 = 0.9999$)

The determination coefficient (r^2) obtained (Table 2) for the regression line demonstrates the excellent relationship between peak area and the concentration of SA, PP and DP.

TABLE 1 Linearity results of the LC method.

components	Concentration (µg/mL)	Equation for regression line	R^2
Sorbic acid	30 – 150	y = 11.576x + 69.486	0.9999
Propylparaben	5 – 30	y = 12.721x - 14.737	0.9998
Domperidone	36 – 180	y = 31.667x - 403.5	0.9999

3.2.2 Precision

The precision of the chromatographic method, reported as percent relative standard deviation (RSD), was estimated by measuring repeatability (intra-day assay precision) on ten replicate injections at 100% test concentration (0.10 mg of SA, 0.02 mg of PP and 0.12 mg of DP/mL). The RSD values for retention time (min) 0.05, 0.09, 0.07%; peak area 0.26, 0.56, 0.19% and peak height were 0.87,

0.62 and 0.78% for SA, PP and DP, respectively. The intermediate precision (inter-day variation) was studied using two LC systems over two consecutive days at three different concentration levels (80, 100, 120 µg/mL for SA, 15, 20, 25 for PP, 90, 120, 150 µg/mL for DP) that cover the assay range (80-120%). Three replicate injections were injected for each solution. The RSD values for both analysts were ≤ 0.67% (Table 3) and illustrated the good precision of this analytical method.

TABLE 3 Results from evaluation of the intermediate precision of the LC method

Concentration (µg/mL)	Day 1, Analyst 1		Day 2, Analyst 2	
	Assay (%)*	%RSD	Assay (%)*	%RSD
Sorbic acid				
80	99.97	0.34	99.95	0.48
100	99.99	0.39	100.03	0.29
120	99.92	0.27	99.93	0.36
Propylparaben				
15	100.01	0.42	99.92	0.56
20	100.11	0.36	99.86	0.37
25	99.98	0.67	100.05	0.23
Domperidone				
90	99.71	0.18	100.02	0.12
120	99.58	0.14	100.13	0.16
150	100.07	0.10	100.23	0.10

*Mean of three replicate

3.2.3 Accuracy

The accuracy was evaluated by the recovery of a known amount of SA, PP and DP in synthetic mixture prepared by mixing SA, PP and DP to placebo, to obtain concentration of 80-120% of normal analytical condition. Calculation of accuracy was carried out as the percentage of drug and preservatives recovered from the synthetic mixture of the drug and preservatives. Mean recovery (Table 4) for

SA, PP and DP from the formulation was between 99.51 and 101.5 ($n = 3$) indicating that the developed method was accurate for the determination of SA, PP and DP in pharmaceutical formulation.

TABLE 4 Results from evaluation of the accuracy of the LC method

Theoretical (% of target level)	Added amount (mg)	Amount founded (mg)	Mean recovery (%)*	RSD (%)
Sorbic acid				
80	2.06	2.10	101.94	0.22
100	4.12	4.14	100.48	0.16
120	6.00	5.98	99.66	0.27
Propylparaben				
80	2.02	2.03	100.49	0.35
100	4.08	4.06	99.51	0.28
120	6.05	6.03	99.67	0.49
Domperidone				
80	2.10	2.13	101.43	0.11
100	4.07	4.09	100.49	0.08
120	6.11	6.16	100.82	0.16

*Mean of three replicate

3.2.4 Specificity

Specificity is the ability of a method to measure analyte response in the presence of its potential impurities. Stress testing of the drug substance can help to identify likely degradation products, which can, in turn, help to establish the degradation pathways and intrinsic stability of the molecule and validate the stability-indicating power of the analytical procedure used. In the present study, injections of the blank were performed to demonstrate the absence of interference with the elution of the SA, PP and DP. These results demonstrate that there was no interference from the other compounds and, therefore, confirm the specificity of the method.

Forced degradation studies were also performed to evaluate the specificity of drug product and each preservative under four stress conditions (heat, UV light, acid, base, oxidative). Solutions of drug and each preservative were exposed to 70°C for 4 h, UV light using a Mineralight UVGL-58 light for 24 h, acidic hydrolysis (0.5 M HCl) for 4 h, basic hydrolysis (0.5 M NaOH) for 4 h and oxidative degradation (3.0% H_2O_2) for 5 h. A summary data of the stress results is shown in Table 5, which showed no changes in retention times of drug and preservative components and no degradation peaks were observed. This was further confirmed by peak purity analysis on a DAD UV detector and, therefore, confirms the specificity of the method.

TABLE 5 Results from evaluation of the forced degradation study of the LC method.

Stress conditions	Sample treatment	Retention time (min)			Assay (%)		
		SA	PP	DP	SA	PP	DP
Reference	Fresh solution	3.88	6.12	8.53	99.97	99.99	100.01
Acidic hydrolysis	0.5 M HCl for 4 h	3.84	6.12	8.52	99.74	99.85	99.98
Basic hydrolysis	0.5 M NaOH for 4 h	3.87	6.11	8.53	99.92	99.67	99.95
Oxidative	3.0% H_2O_2 for 5 h	3.86	609	851	99.87	99.67	99.92
Heat degradation	70 °C for 1 h	3.82	6.10	8.51	99.83	99.57	99.82
Light degradation	UV Light for 24 h	3.85	6.11	8.52	99.97	99.91	99.68

3.2.5 Limits of detection and quantitation

The limit of detection (LOD) and limit of quantitation (LOQ) were determined by the calibration plot method.[22] A specific calibration plot was constructed using samples containing amounts of analytes in the range of LOD and LOQ. The values of LOD and LOQ were 0.14, 0.22 and 5 µg/mL and 25, 5 and 30 µg/mL for SA, PP and DP, respectively, for 20 µL injection volume. LOD and LOQ were calculated by use of the equations:

$$LOD = Cd \times Syx/b$$
$$LOQ = Cq \times Syx/b$$

where Cd and Cq are the coefficients for LOD and LOQ, Syx is the residual variance of the regression, and b is the slope. Calculations were performed by using values of Cd and Cq of 3.3 and 10. Precision at the limits of quantitation and detection was checked by analysis of six test solutions prepared at three levels. The RSD values for peak area were less than 2% for LOQ and less than 5% for LOD solutions.

4. ASSAY RESULTS

Results from analysis of SA, PP and DP pharmaceutical oral suspension products in different batches ($n = 3$) ranged from 99.94 to 99.98%.

5. CONCLUSION

The newly developed LC method is specific, precise, accurate and rapid for the simultaneous determination of sorbic acid, propylparaben and domperidone from pharmaceutical oral suspension formulation. An excellent correlation existed between peak areas and concentration of both drug and preservatives. It is a stability indicating method and suitable for quality control of pharmaceutical preparations containing sorbic acid, propylparaben and domperidone either alone or in combination.

REFERENCES

1. Reynolds, J.E.F. Part 1. Martindale. The extra Pharmacopoeia, 31st ed., Royal Pharmaceutical Society of Great Britan, London, 1996.
2. Shabir, G.A. Determination of combined p-hydroxy benzoic acid preservatives in a liquid pharmaceutical formulation and assay by HPLC. *J. Pharma. Biomed. Anal.* 2004, 34(1), 207-213.
3. Inmaculada, G., Cruz Ortiz, M.; Luis, S.; Carmen, V.; Elisa, G. Advances in methodology for the validation of methods according to the International Organization for Standardization: Application to the determination of benzoic and sorbic acids in soft drinks by high-performance liquid chromatography. *J. Chromatogr. A* 2003, 992(1-2), 11-27.
4. Kuang, L. K.; You-Zung, H. Determination of preservatives in food products by cyclodextrin-modified capillary electrophoresis with multiwavelength detection. *J. Chromatogr. A,* 1997, 768 (2), 334-341.
5. Argekar, A.P.; Shah, S.J. Simultaneous determination of cinnarizine and domepiridone maleate from tablet dosage form by reverse phase ion pair HPLC. *J. Pharm. Biomed. Anal.* 1999, 19(6), 813-817.
6. Veronique, M.; Chantal, S.; Jacques, T. An improved HPLC assay with fluorescence detection for the determination of domperidone and three major metabolites for application to in vitro drug metabolism studies. *J. Chromatogr. B,* 2007, 852(1-2), 611-616.
7. Imran, A.; Gupta, V.K. Prashant Singh and H.V. Pant. Screening of domperidone in wastewater by HPLC and solid phase extraction methods. *Talanta,* 2006, 68(3), 928-931.
8. Koujirou, Y.; Mami, H.; Hajime, K.; Tatsuji, I. Quantitative determination of domperidone in rat plasma by HPLC with fluorescence detection. *J. Chromatogr. B,* 1998, 720(1-2), 251-255.

9. Ittipon, T.; Ranee, S. Analysis of benzoic acid and sorbic acid in Thai rice wines and distillates by solid-phase sorbent extraction and HPLC. *J. Food Comp. Anal.* 2007, 20(3-4), 220-225.

10. Valeria, A.; Lozano, et.al. Simultaneous determination of sorbic and benzoic acids in commercial juices using the PLS-2 multivariate calibration method and validation by HPLC. *Talanta*, 2007, 73(2), 282-286.

11. Masayo, H.; Katsuhiro, I.; Akira, O.; Kazuaki, K. NMR analysis of ion pair formation between timolol and sorbic acid in ophthalmic preparations. *J. Pharm. Biomed. Anal.* 2007, 43(4), 1335-1342.

12. Aluoch-Orwa, J.; Quintens, I.; Roets, E.; Hoogmartens, J. Quantitative analysis of sorbic acid in pharmaceutical cream formulations by liquid chromatography on poly(styrene-divinylbenzene) *Eur. J. Pharm. Sci.*, 1997, 5(3), 155-161.

13. Kaniansky, D.; Masar, M.; Madajova, V.; Marak, J. Determination of sorbic acid in food products by capillary zone electrophoresis in a hydrodynamically closed separation compartment. *J. Chromatogr. A*, 1994, 677(1), 179-185.

14. Giovanni, B.; Pietro, D. Determination of sorbic acid in margarine and butter by HPLC with fluorescence detection. *J. Chromatogr. A*, 1991, 543(1), 69-80.

15. Perez, P.; García, E.; Orriols, A.; Minarro, M. A new validated method for the simultaneous determination of benzocaine, propylparaben and benzyl alcohol in a bioadhesive gel by HPLC. *J. Pharm. Biomed. Anal.* 2005, 39(5), 920-927.

16. Hajkova, R.; Solich, P.; Dvoiak, J. Simultaneous determination of methylparaben, propylparaben, hydrocortisone acetate and its degradation products in a topical cream by RP-HPLC. *J. Pharm. Biomed. Anal.*, 32, (4-5), 2003, 921-927.

17. Seong, K.; Hasuck, K. Simultaneous determination of methylparaben, propylparaben and thimerosal by HPLC and electrochemical detection. *J. Pharm. Biomed. Anal.* 1997, 15(9-10), 1359-1364.

273

18. Diane, K.; Bela, K. Determination of methylparaben, propylparaben and chlorpromazine in chlorpromazine hydrochloride oral solution by HPLC. *J. Chromatogr. B*, 1998, 707(1-2),181-187.

19. Jiri, S.; Marie, P. Separation and determination of ketoprofen, methylparaben and propylparaben in pharmaceutical preparation by micellar electrokinetic chromatography. *J. Pharm. Biomed. Anal.* 2008, 48(2), 452-455.

20. Rafifa, H.; Marie, P.; Andrea, K.; Petr, S.; Jan, S. Separation and determination of clotrimazole, methylparaben and propylparaben in pharmaceutical preparation by micellar electrokinetic chromatography. *J. Pharm. Biomed. Anal.* 2006, 40(1), 215-219.

21. Magnus, A.; Atemnkeng, E.M.; Jacqueline, P. Assay of artemether, methylparaben and propylparaben in a formulated paediatric antimalarial dry suspension. *J. Pharm. Biomed. Anal.* 2007, 43(2), 727-732.

22. International Conference on Harmonization (ICH) of Technical Requirements for the Registration of Pharmaceutical for Human Use, Validation of Analytical Procedures: Text and Methodology Q2(R1), November, 2005.

23. Shabir, G.A. Validation of HPLC methods for pharmaceutical analysis: Understanding the differences and similarities between validation requirements of the U.S. Food and Drug Administration, the U.S. Pharmacopoeia and the International Conference on Harmonization. *J. Chromatogr. A.* 2003, 987(1-2), 57-66.

24. The United States Pharmacopeia-National Formulary (USP–NF), USP 33–NF 28 ed. United States Pharmacopieal Convention, Rockville, Maryland, USA, 2010.

15 SIMULTANEOUS ANALYSIS OF PHENOTHRIN, METHYL-4-HYDROXYBENZOATE AND PROPYL-4-HYDROXYBENZOATE IN HUMAN HEAD LICE MEDICINE BY HPLC

ABSTRACT

A new and simple reversed-phase high-performance liquid chromatographic (HPLC) method has been developed and validated for the simultaneous determination of methyl-4-hydroxybenzoate, propyl-4-hydroxybenzoate and phenothrin in human head lice medicine liquid formulation. The separation was achieved on a Lichrosorb C18, 150 mm × 4.6 mm and 5 µm column with detection wavelength of 254 nm using an isocratic mobile phase mixture of methanol-water (80:20, v/v) at flow rate of 1.0 mL/min. A good linearity correlation ($r^2 \geq 0.9998$) was obtained over the investigated concentration ranges. Recoveries of this method for the three components were from 99.50 to 101.00%. The intra-day and inter-day percent relative standard deviations were less than 2.0%. The proposed method was successfully employed for analysis of methyl-4-hydroxybenzoate, propyl-4-hydroxybenzoate and phenothrin in human head lice medicine liquid products.

Keywords: HPLC, Method validation, Phenothrin, Methyl-4-hydroxybenzoate, Propyl-4-hydroxybenzoate, Head lice medicine.

1. INTRODUCTION

Phenothrin ((3-phenoxyphenyl)methyl 2,2-dimethyl-3-(2-methyl-1-propenyl) cyclopropanecarboxylate)) is a synthetic pyrethroid which has been shown to exhibit high killing activity against houseflies and other household insects, especially highly effective against lice and their eggs [1-2]. Phenothrin (PNT) is used as an active component in pharmaceutical liquid preparations for the treatment of human head lice infections. It is being used widely all over the world for its unique pharmaceutical activity. Some analytical methods are available on the determination of PNT, either alone or in combination with other formulations, such as shampoo, aerosols, dust, emulsifiable concentrates, oils and premixs, by gas chromatography (GC) and high performance liquid chromatography (HPLC) [3-4]. Parabens, a group of alkyl esters of p-hydroxybenzoic acid are widely used as antimicrobial preservatives in cosmetics, food and pharmaceutical products [5]. The parabens are effective over a wide pH range and present a broad spectrum of antimicrobial activity, although they are most effective against yeast and molds. Methyl-4-hydroxybenzoate, (MHB) is also known as Methyl paraben and propyl-4-hydroxybenzoate (PHB) is also known as Propyl paraben (0.05 to 0.25%) have been used for the preservation of various pharmaceutical formulations. Formulator must be fully aware of the procedure for preservative systems in a product need to be analysed to establish their effectiveness throughout shelf life of the product, hence, the simultaneous determination of these preservatives in human head lice medicine liquid products is particularly important both for quality assurance and consumer safety.

Many existing analytical procedures are available in literature for the determination of MHB and PHB preservatives, either alone or in combination with other drugs by HPLC and other techniques [6-20]. Additionally, different analytical methods are available for quantification of PNT. The novelty of the proposed method is the simultaneous analysis of PNT drug and a combination of the two preservatives in human head lice medicine liquid formulation. The objective of this study was to develop and validate a specific, accurate, precise and robust HPLC method for simultaneous determination of PNT, MHB and PHB in human head lice medicine liquid. Figure1 shows chemical structures of the investigated compounds.

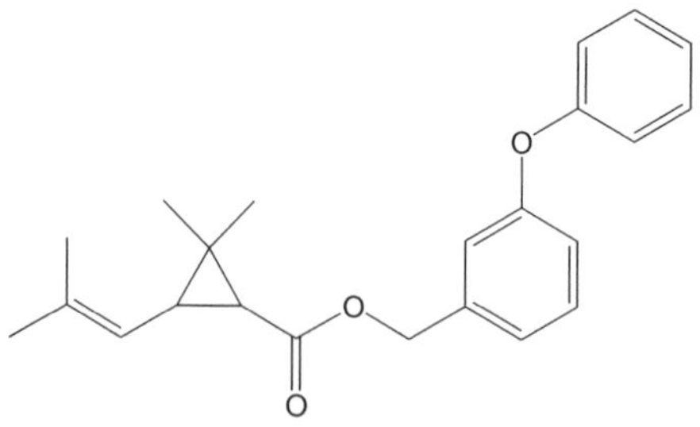

Phenothrin

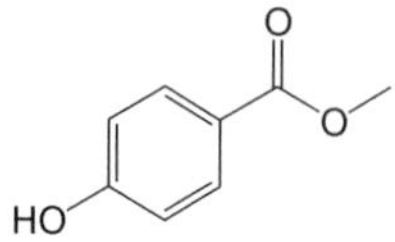

Methyl-4-hydroxybenzoate

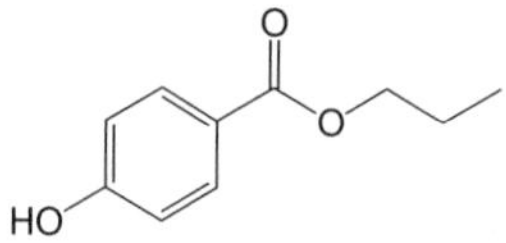

Propyl-4-hydroxybenzoate

FIGURE 1 Chemical structures of the investigated compounds.

2. EXPERIMENTAL

2.1 Chemicals and Reagents

Methanol (HPLC-grade), phenothrin (PNT, analytical standard), methyl-4-hydroxybenzoate (MHB, ≥ 99.0 %) and propyl-4-hydroxybenzoate (PHB, ≥ 99.0 %) were purchased from Sigma-Aldrich (Gillingham, UK). Phenothrin head lice medicine product samples were purchased from local pharmacy (Oxford, UK). Purified water was prepared by Milli-Q system (Bedford, MA, USA).

2.2 HPLC System and Conditions

Chromatographic separation was carried out on a Knauer HPLC system (Berlin, Germany) equipped with a model 1000 LC pump, model 3950 autosampler, model 2600 photodiode-array (PDA) detector and a vacuum degasser was used. The data were acquired via Knauer ClarityChrom Workstation data acquisition software.

The mobile phase consisted of a mixture of methanol and water (80:20, v/v) at a flow rate of 1.0 mL/min. The injection volume was 20 µL and the detection wavelength was set at 254 nm. Reversed-phase LC analysis was performed isocratically at 30°C using a Lichrosorb C_{18} (150 × 4.6 mm, 5 µm) column (Jones Chromatography, Hengoed, UK).

2.3 Standard Preparation

A combined standard solution of MHB, PHB and PNT was prepared in mobile phase, yielding a final concentration of 0.40, 0.25 and 0.35 mg/mL, respectively.

2.4 Sample Preparation

An accurately weighed amount (0.50 g) of liquid sample containing MHB, PHB and PNT components was placed in a 100 mL volumetric flask and dissolved in methanol. 10 mL aliquot solution was added to a 100 mL volumetric flask and diluted in 50 mL methanol and volume made up with mobile phase. The mixture solution was centrifuged at 3500 rpm for 5 min; the clear supernatant portion was collected and filtered through a 0.22 μm membrane filter (Millipore, Watford, UK) and 20 μL of this solution was injected onto the HPLC system.

3. RESULTAS AND DISCUSSION

3.1 Method Development

The procedure for the simultaneous analysis of MHB, PHB and PNT using isocratic HPLC method has been reported. The mobile phase was chosen after several trials with methanol, acetonitrile, water and buffer solutions in various compositions. The best separation was obtained using the mobile phase methanol-water (80:20, v/v). Flow rates between 0.5 and 1.5 mL/min were studied. A flow rate of 1.0 mL/min gave an optimal signal to noise ratio with a reasonable separation time. Using a reversed-phase C_{18} column (150 × 4.6 mm, 5 μm), the retention times for MHB, PHB and PNT were found to be 5.58, 7.12 and 11.18 min with resolution between PHB-MHB 6.03 and PNT-PHB 13.09, respectively. Total time of analysis was less than 12 min. The photodiode-array detector was set at 200 to 400 nm and MHB, PHB and PNT components were extracted at maximum absorption at 254 nm and this wavelength was chosen for the assay method. Typical chromatograms obtained are illustrated in Figure 2 and Figure 3a.

3.1.1 System suitability test

System suitability testing verifies that the HPLC systems are working as expected. It is based on the concept that the equipment, electronics, analytical operations, and samples to be analysed constitute an integral system. System suitability was evaluated and chromatographic parameters calculated from experimental data (n = 6), such as capacity factor (K'), peak asymmetry factor (A), resolution factor (R_s), theoretical plate numbers (N, column efficiency) and relative standard deviation (RSD) of peak areas are given in Table 1. The capacity factor, calculated as a relation of the time of injection of sample to column ($1 < K' < 11$), resolution factor ($R_s > 2$), theoretical plate numbers (N, column efficiency > 2000) and %RSD of peak area (n = 6 < 1.0%) were found to be satisfactory. The values obtained for peak asymmetry factor (A) were significantly lower than theoretical values ($1 < A < 1.5$).

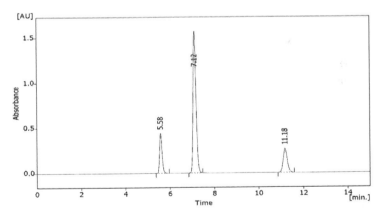

FIGURE 2 LC chromatogram of MHB (t_R 5.58), PHB (t_R 7.12) and PNT (t_R 11.18) obtained from reference standard.

TABLE 1 Results from evaluation of the system suitability parameters of the LC method

Component	Retention time (min)	Peak area (%RSD)	Asymmetry (A)	Capacity (K')	Efficiency (th.pl)	Resolution (R_s)
MHB	5.58	0.13	0.13	4.76	9714	
PHB	7.12	0.09	0.17	2.98	10101	
PNT	11.18	0.16	0.20	10.84	17322	
PHB-MHB						6.03
PNT-PHB						13.09

3.1.2 Robustness

The robustness of an analytical procedure refers to its ability to remain unaffected by small and deliberate variations in method parameter and provides an indication of its reliability for the routine analysis. To determine the robustness of the method the experimental conditions were deliberately altered and retention times (t_R), assay percent, peak tailing, number of theoretical plates and resolutions were evaluated. The mobile phase flow rate was 1.0 mL/min. This was changed by 0.1 units to 0.9 and 1.1 mL/min and the effect was studied. Similarly, the effect of column temperature was studied at 30°C ± 2°C. The effect of mobile phase composition was studied by use of methanol-water 78:22 and 82:18 (v/v). The effect of detection wavelength was studied at 249 and 259 nm. For all changes in conditions the sample was analysed in triplicate. When the effect of altering one set of conditions was tested, the other conditions were held constant at the optimum values. Assay of MHB, PHB and PNT for all deliberate changes of conditions was within 99.4-99.9% (Table 2).

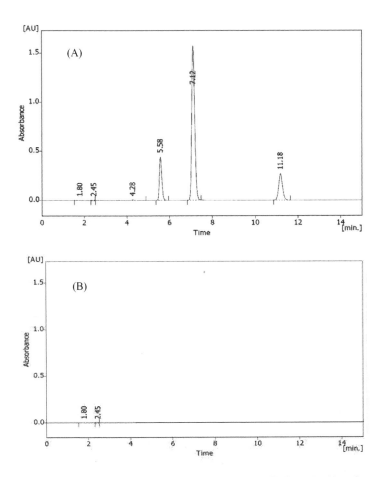

FIGURE 3 LC chromatogram of (A): MHB (t_R 5.58), PHB (t_R 7.12) and PNT (t_R 11.18) obtained from sample, (B): sample matrix blank (placebo).

TABLE 2 Result from evaluation of the robustness study of the LC method

Conditions	t_R (min)	Assay (%)	Peak tailing	Resolution	Theoretical plates
Flow rate (±10% of the optimum flow)					
0.9 mL/min	5.8, 7.1, 11.2	99.8, 99.6, 99.9	1.0, 0.8, 0.9	6.03, 13.9	9714, 10101, 17322
1.1 mL/min	5.6, 7.0, 11.0	99.8, 99.7, 99.8	1.0, 0.7, 0.9	6.02, 13.8	9713, 10104, 17323
Mobile phase composition (±2% of optimum organic modifier concentration					
78 mL	5.9, 7.2, 11.3	99.9, 99.6, 99.8	0.9, 0.8, 1.0	6.03, 13.9	9718, 10102, 17336
82 mL	5.6, 6.9, 10.9	99.6, 98.8, 99.7	1.0, 0.9, 0.7	6.01, 13.7	9719, 10111, 17331
Temperature (±2°C of optimum temperature)					
28°C	5.9, 7.3, 11.3	99.8, 99.5, 99.8	1.0, 0.9, 0.9	6.02, 13.8	9725, 10117, 17327
32°C	5.7, 7.0, 11.0	99.7, 99.8, 99.6	0.9, 0.8, 0.9	6.01, 13.7	9718, 10116, 17341
Wavelength (±5 nm of the optimum wavelength)					
249 nm	5.6, 7.1, 11.2	99.4, 99.8, 99.6	1.2, 0.9, 1.0	6.03, 13,8	9712, 10110, 17334
259 nm	5.6, 7.1, 11.2	99.8, 99.4, 99.8	1.0, 0.8, 0.9	6.03, 13, 9	9712, 10122, 17339

Determinations (n = 3), data for MHB, PHB and PNT, respectively

3.1.3 Stability

The solution stability of MHB, PHB and PNT in the assay method was investigated by leaving sample test solutions in tightly capped volumetric flasks at room temperature for 48 h. The same sample solutions were assayed at 6 h intervals up to the end of the study period against freshly prepared standard solutions. The %RSD of the assay of MHB, PHB and PNT were calculated for the study period during solution stability experiments. The %RSD values of the assay of MHB, PHB and PNT were less than 1.0%. No significant changes were observed during solution stability experiments. The results from these experiments confirm that sample solutions used during assay were stable up to the study period of 48 h.

3.2 Validation of the Method

The developed reversed-phase HPLC method with UV detection was validated with respect to linearity, range, specificity, precision (intermediate precision, repeatability), accuracy, limit of detection, limit of quantitation and stability of analytical solutions following International Conference on Harmonization (ICH) [21], Shabir G.A [22] United States Pharmacopeia (USP) [23] and best practice [24] guidelines.

3.2.1 Linearity and range

The test method ability to obtain a response (i.e. peak area) that is directly proportional to the amount of sample tested was evaluated for all components (MHB, PHB and PNT). Stock standard solutions were prepared and further diluted at six different amounts at six concentration levels from 20% to 160% represents a routine testing amount (0.08-0.64, 0.05-0.40 and 0.07-0.56 mg/mL, MHB, PHB and PNT, respectively. These solutions were injected individually in replicates of three. The mean peak areas were plotted against the corresponding injection amounts. The calibration curves were linear in the range of 20-160% for this assay with correlation coefficients ($r^2 \geq 0.9998$) in each case (Table 3).

TABLE 3 Results from evaluation of the linearity, LOD and LOQ of the LC method

Components	mg/mL	Equation	R^2 value	LOD (mg/mL)	LOQ (mg/mL)
MHB	0.08-0.64	$y = 8571.6x - 106.78$	1.0000	0.03	0.08
PHB	0.05-0.40	$y = 63752x - 401.5$	0.9998	0.02	0.05
PNT	0.07-0.56	$Y = 8704.6x + 105.38$	0.9999	0.04	0.07

3.2.2 Precision

The precision of the chromatographic method, reported as %RSD, was estimated by measuring repeatability (intra-day assay precision) on ten replicate injections at 100% test concentration (0.40 mg of MHB, 0.25 mg of PHB and 0.35 mg of PNT/mL). The %RSD values for retention time (min) 0.09, 0.11, 0.07%; peak area 0.16, 0.13, 0.22% and peak height were 0.42, 0.36 and 0.29% for MHB, PHB and PNT, respectively.

The intermediate precision (inter-day variation) was studied by two analysts over two consecutive days at three different concentration levels (0.32, 0.40, 0.48 mg/mL for MHB, 0.20, 0.25, 0.30 for PHB and 0.28, 0.35, 0.42 mg/mL for PNT) that cover the assay range (80-120%). Three replicate injections were injected for each solution. The %RSD values for peak areas obtained by both analysts were ≤ 1.0% (Figure 4), which illustrated the good precision of this analytical method.

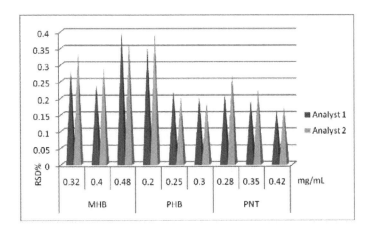

FIGURE 4 Plot for intermediate precision studied by two analysts over two days by LC at three concentrations levels ($n = 3$) of MHB, PHB and PNT.

3.2.3 Accuracy

The accuracy was evaluated by the recovery of a known amount of MHB, PHB and PNT in synthetic mixture prepared by mixing MHB, PHB and PNT to placebo, to obtain concentration of 80-120% of normal analytical condition. Calculation of accuracy was carried out as the percentage of drug and preservatives recovered from the synthetic mixture of the drug and preservatives. Mean recovery (Table 4) for MHB, PHB and PNT from the formulation was between 99.50 and 101.00 ($n = 3$) indicating that the developed method was accurate for the determination of MHB, PHB and PNT in pharmaceutical liquid formulation.

TABLE 4 Results from evaluation of the accuracy of the LC method

Theoretical (% of target level)	Added amount (mg)	Amount found (mg)	Mean recovery (%)*	RSD (%)
Methyl-4-hydroxybenzoate				
80	2.00	2.01	100.05	0.22
100	4.02	4.01	99.75	0.17
120	6.10	6.09	99.84	0.32
Propyl-4-hydroxybenzoate				
80	2.01	2.03	101.00	0.23
100	4.06	4.08	100.49	0.29
120	6.09	6.08	99.84	0.31
Phenothrin				
80	2.02	2.01	99.50	0.16
100	4.04	4.05	100.25	0.15
120	6.06	6.05	99.83	0.26

*Mean of three replicate

3.2.4 Specificity

The ICH guideline defines specificity as the ability to assess unequivocally the analyte in the presence of components that may be expected to be present, such as impurities, degradation products, and matrix components. It is typical to verify lack of any significant contribution to the sample from the matrix via the injection of a suitable "blank" [21]. In the present study, injections of the sample matrix blank (placebo) was analysed to determine the presence of any interfering peaks with the elution of the MHB, PHB and PNT. The results demonstrate that there was no interference observed from the other compounds (Figure 3b).

3.2.5 Limits of detection and quantitation

The limit of detection (LOD) and limit of quantitation (LOQ) were determined by the calibration plot method. A specific calibration plot was constructed using samples containing amounts of analytes in the range of LOD and LOQ. The values of LOD and LOQ were 0.03, 0.02 and 0.04 mg/mL and 0.08, 0.05 and 0.07 mg/mL for MHB, PHB and PNT, respectively, for 20 µL injection volume (Table 3). LOD and LOQ were calculated by using the equations: LOD = $Cd \times Syx/b$ and LOQ = $Cq \times Syx/b$,

where Cd and Cq are the coefficients for LOD and LOQ, Syx is the residual variance of the regression, and b is the slope. Calculations were performed by using values of Cd and Cq of 3.3 and 10. Precision at the LOQ was checked by analysis of six test solutions prepared at three levels. The %RSD values for retention time and peak area was less than 3% for LOQ solutions.

4. CONCLUSION

The isocratic reversed-phase HPLC method with UV detection developed for analysis of MHB, PHB and PNT in human head lice medicine liquid was simple, precise, accurate, specific and robust. The method was critically validated and satisfactory results were obtained for all the method characteristics tested.

The method can be used reliably for routine analysis of formulated product and raw material samples containing MHB, PHB and PNT.

REFERENCES

1. Okuno, Y.; Yamaguchi, T.; Fujita, Y. Botyukagaku, Insecticidal activity of a new synthetic pyrethroidal compound, 3-phenoxy benzyl-(+)cis, trans chrysanthemate (d-phenothrin). *Agric. Biol Chem*. 1976, *41* (1), 42-55.

2. Fujimoto, K.; Itaya, N.; Okuno, Y.; Kadota, T.; Yamaguchi, T. A new insecticidal pyrethroid ester. *Agric. Biol Chem*. 1973, *37* (11), 2681-2682.

3. Sakaue, S.; Kitajima, M.; Horiba, M.; Yamamoto, S. Gas chromatographic and high performance liquid chromatographic determination of *d*-phenothrin (Sumithrin®) in formulations. *Agric. Biol. Chem.* 1981, *45* (5), 1135-1140.

4. Shigeyuki, S.; Megumi, K.; Tadashi, D. High performance liquid chromatographic determination of *d*-phenothrin (Sumithrin®) in shampoo. *Agric. Biol. Chem.* 1985, *49* (9), 2787-2789.

5. Rieger, M.M. In: Weller, A. (Eds.) Handbook of Pharmaceutical Excipients, second ed., The Pharmaceutical press, London, 1994, pp. 310-313. American Pharmaceutical Association, Washington, DC. 1986, 261 pp.

6. Perez, L.P.; Garcia, M.E.; Orriols, A.; Minarro, M.; Tico, J.R.; Sune, J.M. A new validated method for the simultaneous determination of benzocaine, propylparaben, and benzyl alcohol in a bioadhesive gel by HPLC. *J. Pharma. Biomed. Anal.* 2005, *39* (5), 920-927.

7. Sottofatori, E.; Anzaldi, M.; Balbi, A.; Tonello, G. Simultaneous HPLC determination of multiple components in a commercial cosmetic cream. *J. Pharm. Biomed. Anal.* 1998, *18* (1-2), 213–217.

8. Dvorak, J.; Hajkova, R.; Matysova, L.; Novakova, L.; Koupparis, M.A.; Solich, P. Simultaneous HPLC determination of ketoprofen and its degradation products in the presence of preservatives in pharmaceuticals. *J. Pharm. Biomed. Anal.* 2004, *36* (3), 625-629.

9. Popovic, G.; Cakar, M.; Agbaba, D. Determination of bifonazole in creams containing methyl- and propyl p-hydroxybenzoate by derivative spectrophotometric method. *J. Pharm. Biomed. Anal.* 2003, *33* (1), 131-136.

10. Hajkova, R.; Solich, P.; Dvorak, J. Simultaneous determination of methylparaben, propylparaben, hydrocortisone acetate, and its degradation products in a topical cream by RP-HPLC. *J. Pharm. Biomed. Anal.* 2003, *32* (4-5), 921-927.

11. Koundourellis, J. E.; Malliou, E. T.; Broussali, T. A. High performance liquid chromatographic determination of ambroxol in the presence of different

preservatives in pharmaceutical formulations. *J. Pharm. Biomed. Anal.* 2000, *23* (2-3), 469-475.

12. Shabir, G.A.; Lough, W.J.; Shafique, A.A.; Shar, G.Q. Method development and validation for the HPLC assay of phenylformic acid, 2,4-hexadienoic acid, methyl 4-hydroxybenzoate and propyl 4-hydroxybenzoate. *J. Liq. Chromatogr. Relat. Technol.* 2006, *29* (9), 1223-1233.

13. Dalibor, S.; Jitka, H.R.; Ferreira, L.C.; Maria, C.; Montenegro, B.S.M.; Petr, S. Determination of ambroxol hydrochloride, methylparaben and benzoic acid in pharmaceutical preparations based on sequential injection technique coupled with monolithic column. *J. Pharm. Biomed. Anal.* 2006, *40* (2), 287-293.

14. Kokoletsi, M.X.; Kafkala, S.; Tsiaganis, M. A novel gradient HPLC method for simultaneous determination of ranitidine, methylparaben and propylparaben in oral liquid pharmaceutical formulation. *J. Pharm. Biomed. Anal.* 2005, *38* (4), 763-767.

15. Belgaied, J.E.; Trabelsi, H. Determination of cisapride, its oxidation product, propyl and butyl parabens in pharmaceutical dosage form by reversed-phase liquid chromatography. *J. Pharm. Biomed. Anal.* 2003, *33* (5), 991-998.

16. Mahuzier, P.E.; Altria, K. D.; Clark, B.J. Selective and quantitative analysis of 4-hydroxybenzoate preservatives by microemulsion electrokinetic chromatography. *J. Chromatogr. A* 2001, *924* (1-2), 465-470.

17. Lee, M.R., Lin, C.Y.; Li, Z.G.; Tsai, T.F. Simultaneous analysis of antioxidants and preservatives in cosmetics by supercritical fluid extraction combined with liquid chromatography-mass spectrometry. *J. Chromatogr. A* 2006, *1120* (1-2), 244-251.

18. Grosa, G.; Grosso, E.D.; Russo, R.; Allegrone, G. Simultaneous, stability indicating, HPLC-DAD determination of guaifenesin and methyl and propyl-parabens in cough syrup. J. Pharm. Biomed. Anal. 2006, *41* (3), 798-803.

19. Kokoletsi, M.X.; Kafkala, S.; Tsiaganis, M. A novel gradient HPLC method for simultaneous determination of ranitidine, methylparaben and

propylparaben in oral liquid pharmaceutical formulation. *J. Pharm. Biomed. Anal.* 2005, *38*, 991-998.

20. Grosa, G.; Grosso, E.D.; Russo, R.; Allegrone, G. Simultaneous, stability indicating HPLC-DAD determination of guaifenesin and methyl and propyl-parabens in cough syrup. *J. Pharm. Biomed. Anal.* 2006, *41*, 798-803.

21. International Conference on Harmonization (ICH) of Technical Requirements for the Registration of Pharmaceutical for Human Use, Validation of Analytical Procedures: Text and Methodology Q2(R1), November, 2005.

22. Shabir, G.A. Validation of HPLC methods for pharmaceutical analysis: Understanding the differences and similarities between validation requirements of the U.S. Food and Drug Administration, the U.S. Pharmacopoeia and the International Conference on Harmonization. *J. Chromatogr. A.* 2003, *987* (1-2), 57-66.

23. The United States Pharmacopeia-National Formulary (USP–NF), USP 33–NF 28 ed. United States Pharmacopieal Convention, Rockville, Maryland, USA, 2010.

24. Shabir, G.A.; Lough, J.W.; Arain, S.A.; Bradshaw, T.K. Evaluation and application of best practice in analytical method validation. *J. Liq. Chromatogr. Relat. Technol.* 2007, *30* (3), 311-333.

16 DEVELOPMENT AND VALIDATION OF A HPLC METHOD FOR THE DETERMINATION OF 2-PHENOXYETHANOL IN SENSELLE LUBRICANT FORMULATION

ABSTRACT

A simple and rapid reversed-phase liquid chromatographic (RPLC) method was developed and validated for the determination of 2-phenoxyethanol preservative (0.3%, w/w) in senselle lubricant formulation. The separation was achieved with acetonitrile-tetrahydrofuran-water (21:13:66, v/v/v) as mobile phase, a C_8 column, and UV detection at 258 nm. The calibration curve is linear ($r^2 = 0.9999$) from 20-140% of the analytical concentration of 0.75 mg/mL. The mean percent relative standard deviation values for intra- and inter-day precision studies are <1%. The recovery of 2-phenoxyethanol ranged between 99.76 and 100.03% from lubricant formulation. The limit of detection and quantitation are determined to be 0.094 and 0.15 mg/mL, respectively.

Keywords: RP-HPLC, Method validation, 2-Phenoxyethanol, Senselle lubricant formulation

1. INTRODUCTION

2-Phenoxyethanol (ethylene glycol monophenyl ether, $C_8H_{10}O_2$, Figure 1) has been widely used as preservative in cosmetics, skin care products, toiletry, sexual lubricant products and pharmaceutical applications (i.e. in vaccine formulations) [1,2], because of its broad antimicrobial spectrum with good stability and non-volatility [3]. It is a good general bactericide (most active against gram negative bacteria) but a weak fungicide and is generally used in combination with other preservatives. Some liquid chromatographic (LC) methods have been reported for the determination of 2-phenoxyethanol with combination of other components [4, 5]. These reported methods are complicated and time consuming with poor chromatographic separation and longer analytical run time. Furthermore, forced degradation decomposition studies were not included in this work. Determination of 2-phenoxyethanol with solid phase microextraction-gas chromatography-mass spectrometry (SPME-GC-MS/MS) detection has also been reported [6], but this technique is not common use in pharmaceutical quality control laboratories. LC technique has been widely used in pharmaceutical analysis in quality control laboratories because of its sensitivity and specificity. The purpose of this study was to develop and validate a rapid, cost-effective and selective Reversed-Phase Liquid Chromatographic (RPLC) method for routine quality control analysis of 2-phenoxyethanol. Stress testing of the preservative was also conducted, as required by the International Conference on Harmonization (ICH) [7] to support the suitability of the method. As a best practice [8-12] in the subsequent investigation, the new RPLC

294

method was validated according to ICH [7] and U.S. Food and Drug Administration (FDA) [13] guidelines.

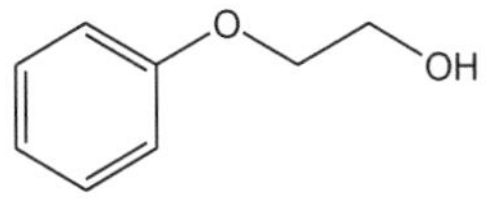

FIGURE 1 Chemical structure of 2-phenoxyethanol.

2. EXPERIMENTAL

2.1 Chemicals and Reagents

Methanol (HPLC-grade), tetrahydrofuran (HPLC-grade), 2-phenoxyethanol (pure > 99%), and formic acid were purchased from Sigma-Aldrich (Gillingham, UK). Distilled water was de-ionised by using a Milli-Q system (Millipore, Bedford, MA).

2.2 LC Instrumentation and Conditions

A Knauer LC system (Berlin, Germany) equipped with a module 1000 LC pump, 3950 autosampler and 2600 photodiode-array (PDA) detector was used. The data were acquired via ClarityChrom data acquisition software. Separation was achieved using Lichrosorb C_8 column (150 × 4.6 mm i.d., 5 μm particle size) from Jones Chromatography (Hengoed, UK). All chromatographic experiments were performed in isocratic mode. The mobile phase consisted of a mixture of acetonitrile-tetrahydrofuran-water (21:13:66, v/v/v) pH 3.0.

The pH was adjusted with formic acid. The flow rate was 1 mL/min, the injection volume was 10 μL, and the temperature was set at 35°C. Chromatograms were recorded at 258 nm using UV detector.

2.3 Standard Preparation

2-Phenoxyethanol standard solutions at 0.75 mg/mL were prepared by dissolving approximately 75 mg of 2-phenoxyethanol in 100 mL mobile phase.

2.4 Sample preparation

An accurately weighed amount (1.25 g) of 2-phenoxyethanol sample lubricant was dissolved in 50 mL mobile phase, yielding a final concentration of 25 mg/mL. The sample was filtered through a sample filtration unit (0.45 μm) and injected into the LC system.

2.5 Calibration Solutions Preparation

Standard solutions of 2-phenoxyethanol were prepared at concentrations of 0.15, 0.45, 0.75, 0.90, and 1.05 mg/mL in the mobile phase and injected in triplicate. Linear regression analysis was carried out on the standard curve generated by plotting the concentration of 2-phenoxyethanol versus peak area response.

3. RESULTS AND DISCUSSION

3.1 Method Development

Efficient chromatography and high sensitivity was achieved by using acetonitrile-tetrahydrofuran-water as the mobile phase with varying detection wavelengths, based on the response of the analyte. However, the analyte peak tailed badly on some C_{18} columns with this mobile phase with longer analytical run time. Using the C_8 column minimised the tailing and shortened the run time. The amount of organic modifier was adjusted so that the assay run time could be reduced for faster analysis of samples. Chromatograms illustrating the separation of 2-phenoxyethanol reference standard and lubricant sample formulation are displayed in Figure 2 and 3, respectively confirming specificity with respect to 2-phenoxyethanol.

The remaining chromatographic conditions listed in Section LC instrumentation and conditions were chosen for the following reasons: the lower flow rate of 1 mL/min was chosen because of the potential problems associated with elevated back pressures. The PDA UV detector was set at 258 nm, λ_{max} for 2-phenoxyethanol. Column temperature was held at 35°C although separation at 30°C and 40°C indicated that slight variation in temperature did not have a significant effect on retention or peak shape. The injection volume of 10 µL and sample concentration of 0.75 mg 2-phenoxyethanol/mL in mobile phase were chosen to simplify sample preparation (further dilution is not needed). This concentration allows purity evaluation.

The peak for 2-phenoxyethanol is well resolved within the linear range for UV detection.

System suitability testing was performed by injecting ten replicate injections of a solution containing 0.75 mg 2-phenoxyethanol/mL. The percent relative standard deviation (RSD) of the peak area responses was measured, giving an average of 0.18 (n = 10). The tailing factor (T) for each 2-phenoxyethanol peak was 1.257, the theoretical plate number (N) was 4803, and the retention time (t_R) variation %RSD was < 1% for ten injections. The RPLC method met these requirements within the accepted limits [9, 13].

For the determination of method robustness within a laboratory during method development a number of chromatographic parameters were determined, which included flow rate, temperature, mobile phase composition, and columns from different lots, In all cases good separation of 2-phenoxyethanol were always achieved, indicating that the method remained selective for 2-phenoxyethanol preservative under the tested conditions.

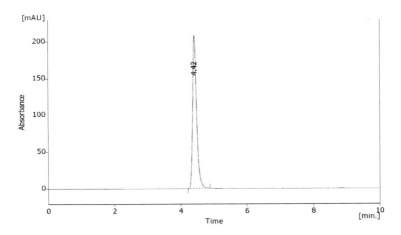

FIGURE 2 RPLC chromatogram of 2-phenoxyethanol standard reference.

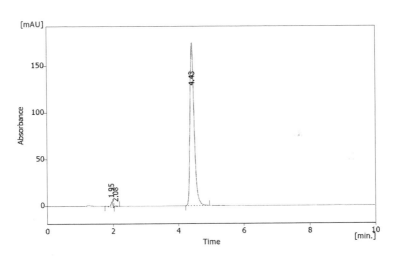

FIGURE 3 RPLC chromatogram of the sample with 0.3% of 2-phenoxyethanol.

3.2 Validation of RPLC Method

3.2.1 Stability of analytical solutions

Samples and standard solutions were chromatographed immediately after preparation and then reassyed after storage at room temperature for 48 h. The results given in Table 1 show that there was no significant change (<1% response factor) in 2-phenoxyethanol concentration over this period.

3.2.2 Linearity

Linearity was studied using five solutions in the concentration range 0.15-1.05 mg/mL (20-140% of the theoretical concentration in the test preparation, $n = 3$). The regression equation was found by plotting the peak area (y) versus the 2-phenoxyethanol concentration (x) expressed in mg/mL. The correlation coefficient (r^2 = 0.9999) obtained for the regression line demonstrates that there is a strong linear relationship between peak area and concentration of 2-phenoxyethanol (Table 1). The analyte response is linear over the range of 80 to120% of the target concentration for 2-phenoxyethanol assay.

TABLE 1 Method validation results of 2-phenoxyethanol preservative

Validation step	Parameters	Concentration (mg/mL)	Results	Acceptance criteria
Standard stability	% change in response factor	0.75	0.08	X < 2
Sample stability	% change in response factor	0.75	0.10	X < 2
Linearity (n = 3; k = 5)	Correlation coefficient (r^2)	0.15-1.05	y = 3238.7x - 380.33 (r^2 = 0.9999)	r^2 = ≥ 0.999
Repeatability (n = 10)	t_R (min) %RSD	0.75	0.07	X < 1
	Peak area %RSD		0.11	
Intermediate precision (n = 6)	Instrument %RSD	0.75	0.22	X < 2
	Analyst %RSD	0.75	0.28	X < 2
System suitability	Peak area %RSD	0.75	0.08	X < 2

3.2.3 Accuracy/recovery studies

The accuracy of the method was evaluated by means of recovery assay, adding known amounts of 2-phenoxyethanol reference standard to a known amount of lubricant formulation in order to obtain three different levels (50%, 100%, and 140%) of addition. The samples were analysed and the mean recovery was calculated. The data presented in Table 2 shows the recovery of 2-phenoxyethanol in spiked samples met the evaluation criteria for accuracy (100 +/- 2.0% over the range of 80 to 120% of target concentration).

TABLE 2 Recovery studies of 2-phenoxyethanol from samples with known concentration

Sample #	Percent of nominal	Amount of 2-phenoxyethanol (mg)		Recovery (%)	RSD (%)
		Added	Recovered		
1	60	3.012	3.013	100.03	0.56
2	60	3.014	3.012	99.93	
3	100	4.022	4.021	99.76	0.24
4	100	4.023	4.017	99.85	
5	140	5.019	5.002	99.66	0.17
6	140	5.022	5.018	99.92	
Mean				99.86	

3.2.4 Precision

The precision of the method was investigated with respect to repeatability (intra-day precision) and intermediate precision (inter-day variation). Repeatability of the method was evaluated by assaying six replicate injections of the 2-phenoxyethanol at 100% of test concentration (0.75 mg/mL). The %RSD of the retention time (min) and relative percent peak area were found to be less than 0.12% (Table 1). Intermediate precision (inter-day variation) was demonstrated by two analysts using two LC systems and evaluating the relative peak area percent data across the two LC systems at three concentration levels (60, 100, and 120%) that cover the assay method range (0.15-1.05 mg/mL). the mean and %RSD across the systems and analysts were calculated from the individual relative percent peak area mean values at the 50%, 100%, and 125% of the test concentration. The %RSD values for both instruments and analysts were ≤ 0.28% (Table 1) and illustrated good precision of RPLC method.

3.2.5 Specificity/selectivity

The RPLC-PDA/UV isoplot chromatogram (Figure 4) demonstrates a good separation of the 2-phenoxyethanol. The isoplot chromatogram data consist of PDA UV-vis absorption spectra from 200 to 300 nm for each point along the chromatogram. Injections of the extracted placebo were also performed to demonstrate the absence of interference with the elution of the 2-phenoxyethanol. These results demonstrate (Figure 5) that there was no interference from the other materials in the lubricant formulation and, therefore confirm the specificity of the RPLC method.

Forced degradation studies were performed to evaluate the specificity of 2-phenoxyethanol under four stress conditions (heat, UV light, acid, base). Solutions of 2-phenoxyethanol were exposed to 60°C for 1 h, UV light using a UVL-56 lamp for 24 h, acid (1 M HCl) for 24 h and base (1 M NaOH) for 4 h. A summary of the stress results (retention time (t_R), peak area, resolution (R) and theoretical plate numbers, (N) is shown in Table 3. Under acid (major degradation) and alkaline (minor degradation) hydrolysis conditions, the 2-phenoxyethanol content decreased and additional peaks were observed (Figure 6). No degradation was observed under other hydrolysis conditions (heat, UV light) studied. The addition peak detected at 1.65 min under acid and 1.95 min under alkaline conditions. This was further confirmed by peak purity analysis on a PDA UV detector. The 2-phenoxyethanol analyte obtained by acid hydrolysis was well resolved from the additional peak (Figure 7), indicating the specificity of the method.

TABLE 3 Force degradation studies data for 2-phenoxyethanol preservative

Stress conditions	Sample treatment	t_R (min)	Area (mAU s)	R	N	Degradation
Reference	Fresh solution	4.417	1854.90	-	4803	No degradation
Heat	50 °C for 1 h	4.417	1754.25	-	6079	No degradation
Light	UV Light for 24 h	4.417	1808.36	-	4803	No degradation
Acid	1M HCl for 24 h	4.433	1590.07	16.42	6125	Degradation
		1.650	762.36	-	3394	
Base	1M NaOH for 4 h	4.433	1538.41	2.64	6125	Degradation
		1.950	66.15	-	3033	

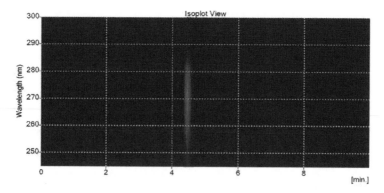

FIGURE 4 RPLC-PDA/UV isoplot chromatogram of 2-phenoxyethanol.

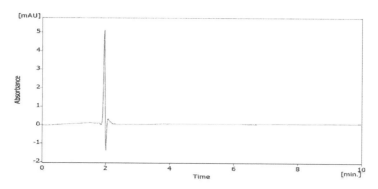

FIGURE 5 RPLC chromatogram of placebo.

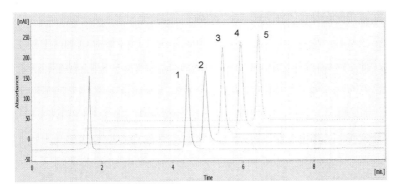

FIGURE 6 RPLC chromatograms of hydrolysed 2-phenoxyethanol under stress conditions (1) acid; (2) base; (3) heat at 60°C; (4) reference standard; (5) UV light.

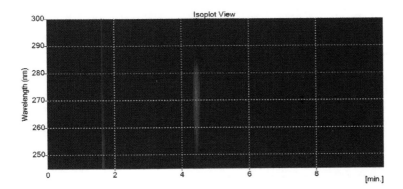

FIGURE 7 RPLC-PDA/UV isoplot chromatogram of hydrolysed 2-phenoxyethanol under acedic stress condition. Degradation peak eluted at 1.65 min.

3.2.6 Limits of detection and quantitation

The limit of detection (LOD) and limit of quantitation (LOQ) of 2-phenoxyethanol was determined based on standard deviation (σ) of response and slope (s). 2-Phenoxyethanol solutions were prepared in the range 0.05-250 µg/mL and injected in triplicate. Average peak area of analyte was plotted against concentration. LOD and LOQ were calculated by using the following equations:

LOD = (3.3 σ)/s   (Eq.1)
LOQ = (10 σ)/s   (Eq.2)

The LOD was determined to be 0.095 mg/mL and LOQ was found to be 0.15 mg/mL for 2-phenoxyethanol with %RSD less than 0.14% for six replicate injections.

3.2.7 System suitability test

A system suitability test was performed to determine the accuracy and precision of the system by injecting six replicate injections of 2-phenoxyethanol standard solution. The RSD of the peak areas responses was measured. The RSD for 2-phenoxyethanol was 0.08% as can be seen in Table 1.

4. CONCLUSION

A simple and rapid reversed-phase liquid chromatographic method with UV spectrophotometer detection was developed for the determination of 2-phenoxyethanol in senselle lubricant formulation.

The method was validated and the results obtained were accurate and precise with RSD < 1% in all cases and no significant interfering peaks were detected. The method is specific, selective, robust and reliable for routine use in quality control for analysis of 2-phenoxyethanol in bulk senselle lubricant sample, raw materials, and final products release.

REFERENCES

1. Cecilia, A.; Anna, E.; Marco, A.; Marisanna, C.; Paolo, N.; Anna, D.S. *In vitro* induction of apoptosis vs. necrosis by widely used preservatives: 2-phenoxyethanol, a mixture of isothiazolinones, imidazolidinyl urea and 1,2-pentanediol. *Biomed. Pharmacol.* 2002, 63, 437-453.

2. Hall, A.L. Cosmetically acceptable phenoxyethanol. In: Kabara, J.J. editor, Cosmetics and drug preservation, principles and practice. New York, Marcel Dekker, 1984, p. 79-108.

3. The Japanese standards of cosmetic ingredients-with commentary, second ed., The Society of Japanese Pharmacopoeia, Yakuginippousha, Tokyo, 1984.

4. Borremans, M.; Van, L.J.; Roos, P.; Goeyens; L. Validation of HPLC analysis of 2-phenoxyethanol, 1-phenoxypropan-2-ol, methyl, ethyl, propyl, butyl and benzyl 4-hydroxibenzoate (parabens) in cosmetic products, with emphasis on decision limit and detection capability. *Chromatographia*, 2004, 59, 47-53.

5. Sharma, B.; Joseph, A.; Sood, A. A simple and rapid method for quantifying 2-phenoxyethanol in diphtheria, tetanus and w-pertussis vaccine. *Biologicals*, 2008, 36, 61-63.

6. Eva, K.; Kateřina, R.; Jana, H.; Jan, P.; Jitka, K.; Vladimír, K. Development of an SPME–GC–MS/MS procedure for the monitoring of 2-phenoxyethanol in anaesthetised fish. *Talanta*, 2008, 75, 1082-1088.

7. International Conference on Harmonization (ICH), Q2(R1): Validation of analytical procedures: Text and Methodology, 2005.

8. Shabir, G.A.; Lough, W.J.; Shafique, A.A.; Bradshaw, T.K. Evaluation and application of best practice in analytical method validation. *J. Liq. Chromatogr. Rel. Technol.* 2007, 30 (3), 311-333.

9. Shabir, G.A. Validation of HPLC methods for pharmaceutical analysis: Understanding the differences and similarities between validation requirements of the U.S. Food and Drug Administration, the U.S. Pharmacopoeia and the International Conference on Harmonization. *J. Chromatogr. A.* 2003, 987(1-2), 57-66.

10. Shabir, G.A. Step-by-step analytical methods and protocol in the quality system compliance industry. *J. Validation Technol.* 2004, 10(4), 314-324.

11. Shabir, G. A. A practical approach to validation of HPLC methods under current good manufacturing practices. *J. Validation Technol.* 2004, 10(3), 210-218.

12. Shabir, G. A. HPLC method development and validation for pharmaceutical analysis. *Pharma. Technol. Eur.*, 2004, 16(3), 37-49.

13. Reviewer Guidance: Validation of Chromatographic Methods, Food and Drug Administration (FDA), Center for Drug Evaluation and Research (CDER), 1994.

17 SIMULTANEOUS DETERMINATION OF *P*-HYDROXYBENZOIC ACID, 2-PHENOXYETHANOL, METHYL-*P*-HYDROXYBENZOATE, ETHYL-*P*-HYDROXYBENZOATE, PROPYL-*P*-HYDROXYBENZOATE, ISO-BUTYL-*P*-HYDROXYBENZOATE AND N-BUTYL-*P*-HYDROXYBENZOATE IN SENSELLE LUBRICANT FORMULATION USING HPLC

ABSTRACT

A new and simple reversed-phase high-performance liquid chromatographic (HPLC) method has been developed and validated for the simultaneous determination of *p*-hydroxybenzoic acid, 2-phenoxyethanol, methyl-*p*-hydroxybenzoate, ethyl-*p*-hydroxybenzoate, propyl-*p*-hydroxybenzoate, iso-butyl-*p*-hydroxybenzoate and n-butyl-*p*-hydroxybenzoate preservatives in senselle lubricant formulation. The seven compounds were separated on a C_{18} ChromSpher polymeric octadecylsilane (ODS)-encapsulated spherical silica column with acetonitrile-tetrahydrofuran-water, 22:14:64 (v/v/v) as mobile phase at flow rate of 1.5 mL/min. The analysis was performed with ultraviolet (UV) detection at 258 nm. The analysis time was <15 min. The method was validated with respect to linearity, precision, accuracy, selectivity, specificity and limits of detection and quantitation. The calibration curves showed good linearity over the concentration range of 5.4–217 µg/mL. The correlation coefficients were ≥ 0.9996 in each case. The relative standard deviation (RSD) values for intra- and inter-day precision studies were <1%. The procedure describe here is simple, selective and is suitable for routine quality control analysis and stability tests.

Keywords: HPLC, validation, p-hydroxybenzoic acid, 2-phenoxyethanol, methyl-p-hydroxybenzoate, ethyl-p-hydroxybenzoate, propyl-p-hydroxybenzoate, iso-butyl-p-hydroxybenzoate, n-butyl-p-hydroxybenzoate, senselle lubricant formulation

1. INTRODUCTION

Senselle is a proprietary non-hormonal water-based preparation used as a vaginal lubricant to relieve dryness of the vagina, which can occur at the menopause. Senselle is freely available over the counter in the form of a liquid. p-Hydroxybenzoic acid (HBA), 2-phenoxyethanol (POE), methyl-p-hydroxybenzoate (MHB), ethyl-p-hydroxybenzoate (EHB), propyl-p-hydroxybenzoate (PHB), iso-butyl-p-hydroxybenzoate (IBH) and n-butyl-p-hydroxybenzoate (NBH) have been widely used as antimicrobial and anti-fungal agents as single preservative but more often in combinations in food, beverage, cosmetics, senselle lubricant products and pharmaceutical formulations[1,2]. These preservatives are commonly added to these formulations due to their broad antimicrobial spectrum with good stability and non-volatility [3] to prevent alteration and degradation of the product formulations and to prolong product shelf life. Industry must be fully aware of the procedure for preservative systems in a product need to be analysed to establish their effectiveness throughout shelf life of the product. Hence, the simultaneous determination of these preservatives in commercial pharmaceutical products is particularly important both for quality assurance and consumer safety. Many existing analytical procedures are available in literature for the determination of present preservatives studied, either alone or in combination with other drugs by high-performance liquid chromatography (HPLC) and other techniques [4-13]. Such a method is important as there seem to be an increasing trend in using combination of preservatives, not only in pharmaceutical formulations but also in food industry, cosmetic and senselle lubricant products. Moreover, many of the reported methods use complicated and labour-intensive pre-treatment procedures such as steam distillation multiple-steps, gradient elution and solid phase extractions. Also no report is available on these seven preservatives analysing together in single analytical run. Therefore, the purpose of the present study was to develop and validate a new, simple, accurate, and robust isocratic reversed-phase HPLC method for the determination of HBA, POE, MHB, EHB, PHB, IBH and NBH preservatives in a single chromatographic run suitable for preservatives raw materials, bulk samples and finished products release. Validating analytical method is a crucial part of successful product development, testing and quality. The determination of preservatives both alone or in formulated products is important and provides a difficult analytical challenge. As a best practice [14], in the subsequent investigation, the new reversed-phase HPLC assay method was validated according to criteria described in the literature [15-20]. The developed and validated method was applied to the analysis of these preservatives in commercially available senselle lubricant products.

2. EXPERIMENTAL

2.1 Materials

HPLC-grade acetonitrile and tetrahydrofuran (THF) were obtained from VWR International (Poole, UK). p-Hydroxybenzoic acid, 2-phenoxyethanol, methyl-p-hydroxybenzoate, ethyl-p-hydroxybenzoate, propyl-p-hydroxybenzoate, iso-butyl-p-hydroxybenzoate, n-butyl-p-hydroxybenzoate and formic acid were purchased from Sigma-Aldrich Chemicals (Gillingham, UK). All chemicals were analytical grade. Distilled water was de-ionised by using a Millipore Milli-Q system (Watford, UK).

2.2 HPLC instrumentation and conditions

A Knauer (Berlin, Germany) HPLC system equipped with a pump model 1000, autosampler model 3950, photodiode-array (PDA) detector model 2600 and a vacuum degasser, all controlled by a ClarityChrom software, was used. The chromatographic separation was achieved isocratically at 30°C using a 250 mm × 4.6 mm, C_{18} ChromSpher polymeric octadecylsilane (ODS)-encapsulated spherical silica column with a 5 mm particle size (Varian, UK). The mobile phase consisted of a mixture of acetonitrile-tetrahydrofuran-water (22:14:64, v/v/v) adjusted to pH 3.0 ±0.05 with formic acid was used. The flow rate was 1.5 mL/min and injection volume was 10 µL. The eluent was monitored with a UV detector set at 258 nm.

2.3 Standard preparation

A combined standard stock solution of HBA, POE, MHB, EHB, PHB, IBH AND NBH was prepared in mobile phase yielding a final concentration of 0.025, 0.75, 0.145, 0.035, 0.018, 0.018 and 0.035 mg/mL, respectively.

2.4 Sample preparation

An accurately weighed amount (1.25 g) of sample in a 100 mL volumetric flask and 50 mL of mobile phase was added; the mixture was extracted in the ultrasonic bath for 5 min at room temperature and diluted with mobile phase to the mark. The solution was filtered through 0.45 µm membrane filter and 10 µL was injected.

3. RESULTS AND DISCUSSION

3.1 Method Development

The procedure for the simultaneous analysis of seven preservatives HBA, POE, MHB, EHB, PHB, IBH and NBH using isocratic reversed-phase HPLC method has been reported. The mobile phase was chosen after several trials with methanol, acetonitrile, tetrahydrofuran, water and buffer solutions in various compositions and at different pH values. The best separation was obtained using the mobile phase acetonitrile-tetrahydrofuran-water (22:14:64, v/v/v) adjusted to pH 3.0 ±0.05 with formic acid. Flow rates between 1.0 and 2.0 mL/min were studied. A flow rate of 1.5 mL/min gave an optimal signal to noise ratio with excellent separation time. The photodiode-array detector was set at 190 to 400 nm and HBA, POE, MHB, EHB, PHB, IBH and NBH preservatives were extracted at maximum absorption at 258 nm and this wavelength was chosen for the assay method. Using a reversed-phase 250 mm × 4.6 mm, C_{18} ChromSpher polymeric octadecylsilane (ODS)-encapsulated spherical silica column with a 5 mm particle size, the retention times for HBA, POE, MHB, EHB, PHB, IBH and NBH were found to be 4.62, 6.29, 8.28, 9.05, 10.34, 11.55 and 12.80 min, respectively. Total time of analysis was < 15 min. Using these optimized conditions, typical chromatogram obtained from test sample is illustrated in Figure 1a.

Robustness verification studies were also performed in the method development phase. The robustness of an analytical method is defined as the measure of its capacity to remain unaffected by small but deliberate variations in the method parameters and provides an indication of its reliability during normal usage. One way to gauge robustness is to examine some relevant factors, which might influence the reliability of the developed method. Selected factors, namely the mobile phase composition (±2 mL), flow rate (±0.1 unit), temperature (±2°C), wavelength (±5 nm) and column from different lots were investigated. In all cases, good separations of all seven preservatives were always achieved, indicating that the analytical method remained selective and robust for HBA, POE, MHB, EHB, PHB, IBH and NBH components under the optimized conditions.

System suitability testing verifies that the HPLC system is working as expected. It is based on the concept that the equipment, electronics, analytical operations and samples to be analysed constitute an integral system. System suitability was evaluated by injecting solution of HBA, POE, MHB, EHB, PHB, IBH and NBH components (concentration of 0.025, 0.75, 0.145, 0.035, 0.018, 0.018 and 0.035 mg/mL), respectively, in six replicates at the beginning of the validation run. System suitability parameters calculated from chromatogram (Fig. 1a), such as peak tailing factor (T), resolution factor (R_s), theoretical plate numbers (N, column efficiency) and coefficient of variation (%CV) of peak areas are given in Table 1. The calculated resolution factor ($R_s > 2$), theoretical plate numbers (N, column efficiency > 2000) and %RSD of peak area ($n = 6 < 0.22\%$) were found to be satisfactory.

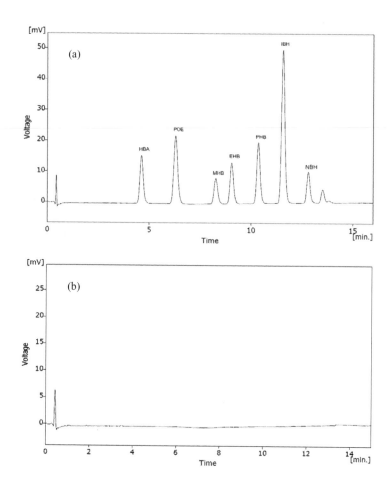

FIGURE 1 HPLC chromatogram obtained for: (a) test sample of *p*-hydroxybenzoic acid (HBA), 2-phenoxyethanol (POE), methyl-*p*-hydroxybenzoate (MHB), ethyl-*p*-hydroxybenzoate (EHB), propyl-p-hydroxybenzoate (PHB), iso-butyl-*p*-hydroxybenzoate (IBH) and n-butyl-*p*-hydroxybenzoate (NBH) preservatives, (b) placebo sample without adding preservatives in lubricant formulation.

TABLE 1 System Suitability Results Summary of the HPLC Method for the Assay of Preservatives

Analyte	t_R (min)	Area (%CV)*	T	N	R_s
p-Hydroxybenzoic acid	4.62	0.21	1.31	4261	-
2-Phenoxyethanol	6.29	0.15	1.30	5053	5.27
Methyl-p-hydroxybenzoate	8.28	0.11	1.21	11287	5.98
Ethyl-p-hydroxybenzoate	9.05	0.17	1.29	16335	2.62
Propyl-p-hydroxybenzoate	10.34	0.09	1.34	21330	4.57
iso-Butyl-p-hydroxybenzoate	11.55	0.22	1.23	26606	4.28
n-Butyl-p-hydroxybenzoate	12.80	0.16	1.31	36206	4.54

*% Coefficient of variation (%CV) as % relative standard deviation (%RSD) are both terms describe the same statistical operation, $n = 6$.

3.2 Method Validation

3.2.1 Linearity and range

The test method ability to obtain a response (i.e. peak area) that is directly proportional to the amount of sample tested was evaluated for all components (HBA, POE, MHB, EHB, PHB, IBH and NBH). Stock standard solutions were prepared and further diluted at six different amounts 30%, 60%, 80%, 100%, 120% and 150% represents a routine testing amount (0.025, 0.75, 0.145, 0.035, 0.018, 0.018 and 0.035 mg/mL, HBA, POE, MHB, EHB, PHB, IBH and NBH, respectively) were injected in replicates of three. The mean peak areas were plotted against the corresponding injection amounts. The calibration curves were linear in the range of 30-150% for this assay with correlation coefficients ($r^2 \geq 0.9996$) in each case (Table 2).

TABLE 2 Linearity Results Summary of the HPLC Method for the Assay of Preservatives

Analyte	Range (µg/mL)	Equation for Regression Line	R^2 Value[*]
p-Hydroxybenzoic acid	7.5-37.5	y = 9.7393x - 49.384	0.9999
2-Phenoxyethanol	225-1125	y = 0.6615x - 138.6	1.0000
Methyl-p-hydroxybenzoate	43.5-217.5	y = 3.204x - 122.81	0.9998
Ethyl-p-hydroxybenzoate	10.5-52.5	y = 4.4139x + 26.661	0.9999
Propyl-p-hydroxybenzoate	5.4-27	y = 13.435x - 7.4117	0.9998
iso-Butyl-p-hydroxybenzoate	5.4-27	y = 13.58x - 11.471	0.9996
n-Butyl-p-hydroxybenzoate	10.5-52.5	y = 4.3077x + 34.15	0.9998

*Acceptance criteria: R^2 value not less than 0.99

3.2.2 Repeatability and intermediate precision

The precision of the chromatographic method, reported as %RSD, was estimated by measuring repeatability (intra-day assay precision) on six replicate injections at 100% test concentration (0.025, 0.75, 0.145, 0.035, 0.018, 0.018 and 0.035 mg/mL) for HBA, POE, MHB, EHB, PHB, IBH and NBH, respectively (n = 6). The %RSD values for retention times (min) < 0.09%, peak area < 0.23% and peak relative percent purity (% area) were found to be < 0.18% for each preservative studied.

The intermediate precision (within-laboratory variation) was studied by two analysts over two consecutive days at three different concentration levels 20, 25, 30 for HBA, 600, 750, 900 for POE, 116, 145, 174 for MHB, 28, 35, 42 for EHB and NBH, 14.4, 18, 21.6 µg/mL for PHB and IBH that cover the assay range (80-120%). Three replicate injections were injected for each solution. The mean and %RSD across the two analysts were calculated from the individual relative % purity mean values at the 80%, 100% and 120% testing amounts.

The %RSD values for peak areas obtained by both analysts were ≤ 1.0% (Figure 2), and met the intermediate precision criteria (RSD < 2.0%) which illustrated the good precision of this analytical method.

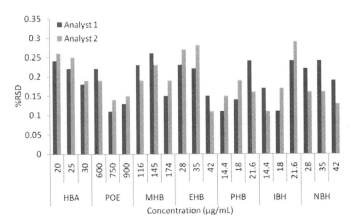

FIGURE 2 Comparison (%RSD) of intermediate precision evaluated by two analysts at three different concentration levels of drug analytes (n = 3). Error bars represent ±1SD

3.2.3 Accuracy

Accuracy is the measurement of how close the experimental value is to the true value. Accuracy/recovery study was investigated at three sample testing amount (80%, 100%, and 120% of the routine testing amount) of the test method range. ICH guideline recommend assessing accuracy using a minimum of nine determinations over a minimum of three concentration levels covering the specific range (e.g., 3 concentrations/3 replicates each of the total analytical procedure) [18]. The accuracy of the method was evaluated by spiking known amount of HBA, POE, MHB, EHB, PHB, IBH and NBH in the formulation (placebo) at 80 %, 100% and 120% level (n = 3) of that in a test preparation and injected in triplicate. The amount of analytes added was calculated for percent recovery and coefficient of variation (%CV). Accuracy acceptance criteria (98-102%) were met at each of the three amounts (Table 3).

TABLE 3 Accuracy/Recovery Results Summary of the HPLC Method for the Assay of Preservatives

Analyte	Applied Concentration (Percent of Target) (n = 3)		
	80 (%)	100 (%)	120 (%)
p-Hydroxybenzoic acid	99.49* (±0.26)**	99.74 (±0.18)	101.06 (±0.56)
2-Phenoxyethanol	99.93 (±0.17)	99.97 (±0.34)	99.27 (±0.23)
Methyl-p-hydroxybenzoate	98.94 (±0.25)	99.91 (±0.21)	100.15 (±0.62)
Ethyl-p-hydroxybenzoate	98.55 (±0.29)	99.89 (±0.19)	100.23 (±0.19)
Propyl-p-hydroxybenzoate	101.02 (±0.35)	99.37 (±0.30)	99.60 (±0.38)
iso-Butyl-p-hydroxybenzoate	99.65 (±0.32)	99.87 (±0.17)	99.28 (±0.46)
n-Butyl-p-hydroxybenzoate	98.93 (±0.46)	99.18 (±0.46)	102.02 (±0.54)

*%Recovery, **% Coefficient of variation (%CV) as % relative standard deviation (%RSD) are both terms describe the same statistical operation, n = 3.

3.2.4 Specificity

Injections of the blank were performed to demonstrate the absence of interference with the elution of HBA, POE, MHB, EHB, PHB, IBH and NBH preservatives. These results demonstrate (Figure 1b) that there was no interference from the other compounds and, therefore, confirms the specificity of the method.

3.2.5 Limits of detection and quantitation

Limit of detection (LOD) is the lowest amount of analyte in a sample that can be detected, but not necessary quantitated as an exact value. Limit of quantitation (LOQ) is the lowest amount of analyte in a sample that can be determined with acceptable precession and accuracy under the stated experimental conditions. LOD and LOQ were determined by the calibration plot method. A specific calibration plot was constructed using samples containing amounts of analytes in the range of LOD and LOQ. The values of LOD and LOQ were 2.4, 22, 24, 5.4, 3.5, 4.5 and 5.4 µg/mL and 7.5, 225, 43.5, 10.5, 5.4, 5.4 and 10.5 µg/mL for HBA, POE, MHB, EHB, PHB, IBH and NBH, respectively, for 10 µL injection volume. LOD and LOQ were calculated using the following equations:

$$LOD = Cd \times Syx/b \text{ and } LOQ = Cq \times Syx/b$$

where Cd and Cq are the coefficients for LOD and LOQ, Syx is the residual variance of the regression, and b is the slope. Calculations were performed by using values of Cd and Cq of 3.3 and 10. Precision at the limits of quantitation and detection was checked by analysis of six test solutions prepared at three levels. The %RSD values for peak area were < 5% for LOQ solutions (Table 2), which indicates the sensitivity of the method is adequate.

3.2.6 Stability

The senselle lubricant test sample stability in the autosampler was investigated on HPLC system. Test sample solutions was prepared at 100% test concentration and placed in the autosampler at ambient room temperature (15 – 22°C) and injected. After completion of the analysis run, the samples were remains in the autosampler at room temperature for 48 hours and then again analysed. A summary of stability results is given in Table 4. The difference in relative %purity value (%area) between the first injection of test sample and the last injection was less than 2%.

This value was comparable to the variability of the test method and revealed no tendency of sample degradation, therefore, the senselle lubricant test sample was considered stable in the autosampler for at least 48 hours at ambient room temperature. This is beyond the routine time occurred in the final standard testing procedure.

4. APPLICATION OF THE METHOD

To demonstrate the applicability of the present method, commercially available three batches of senselle lubricant products containing HBA, POE, MHB, EHB, PHB, IBH and NBH preservatives were analysed. Assay results for three samples of senselle lubricant expressed as the percentage of the label claim, were found between 99.26 to 100.11%. Results showed that the content of HBA, POE, MHB, EHB, PHB, IBH and NBH preservatives in the senselle lubricant formulation was to the counter requirements (90-110% of the label claim). The chromatogram obtained from sample analysis is given in Figure 1a. The above results demonstrated that the developed method achieved rapid and accurate determination of preservatives studied and can be used for the simultaneous determination of HBA, POE, MHB, EHB, PHB, IBH and NBH preservatives in drug substances and senselle lubricant formulations.

TABLE 4 Stability Results Summary of the HPLC Method for the Assay of Preservatives

Analyte	Cumulative Time	Relative %Purity Value	%Difference ($n = 3$)
First Injection-Start	0 Hour	N/A	N/A
p-Hydroxybenzoic acid		99.84	0.26
2-Phenoxyethanol		99.93	0.47
Methyl-p-hydroxybenzoate		99.96	0.77
Ethyl-p-hydroxybenzoate		99.47	0.40
Propyl-p-hydroxybenzoate		99.57	0.41
iso-Butyl-p-hydroxybenzoate		99.54	0.71
n-Butyl-p-hydroxybenzoate		99.85	1.56
Last Injection-Start	48 Hours	-	-
p-Hydroxybenzoic acid		99.58	0.28
2-Phenoxyethanol		99.46	0.57
Methyl-p-hydroxybenzoate		99.19	0.45
Ethyl-p-hydroxybenzoate		99.07	0.79
Propyl-p-hydroxybenzoate		99.16	0.44
iso-Butyl-p-hydroxybenzoate		98.83	0.94
n-Butyl-p-hydroxybenzoate		98.29	0.83

Shaded areas do not require data to be entered.

5. CONCLUSION

The reversed-phase LC method with UV detection developed for analysis of p-hydroxybenzoic acid, 2-phenoxyethanol, methyl-p-hydroxybenzoate, ethyl-p-hydroxybenzoate, propyl-p-hydroxybenzoate, iso-butyl-p-hydroxybenzoate and n-butyl-p-hydroxybenzoate preservatives is rapid, simple, precise, accurate, specific and robust. The method was critically validated and satisfactory results were obtained for all the method validation performance characteristics tested. The method is stability-indicating and can be conveniently applied for the testing of preservatives raw materials used in formulations and batch release by industry.

REFERENCES

1. Rieger, M.M, in: Wade, A.; Weller, P.J. (eds.), Handbook of Pharmaceutical Excipients, second ed., The Pharmaceutical Press, London, 1994; pp. 310-313.
2. Shabir, G.A. Determination of combined p-hydroxy benzoic acid preservatives in a liquid pharmaceutical formulation and assay by HPLC. J. Pharm. Biomed. Anal. 2004, 34 (1), 207-13.
3. The Japanese standards of cosmetic ingredients-with commentary, second ed., The Society of Japanese Pharmacopoeia, Yakuginippousha, Tokyo; 1984.
4. Shabir, G.A. Method development and validation of preservatives determinations (benzyl alcohol, ethylene glycol monophenyl ether, methyl hydroxybenzoate, ethyl hydroxybenzoate, propyl hydroxybenzoate and butyl hydroxybenzoate) using HPLC. J. Liq. Chromatogr. Relat. Technol. 2007, 30 (13), 1951-1962.
5. Shabir, G.A.; Lough, W.J.; Arain, S.A.; Shar, G.Q. Method development and validation of preservatives (phenylformic acid, 2,4-hexadienoic acid, methyl 4-hydroxybenzoate, and propyl 4-hydroxybenzoate) by HPLC. J. Liq. Chromatogr. Relat. Technol. 2006, 29 (9), 1223-1233.
6. Sottofatori, E.; Anzaldi, M.; Balbi, A.; Tonello, G. Simultaneous HPLC determination of multiple components in a commercial cosmetic cream. J. Pharm. Biomed. Anal. 1998,18 (1), 213-217.
7. Hajkova, R.; Solich, P.; Dvorak, J. Simultaneous determination of methylparaben, propylparaben, hydrocortisone acetate, and its degradation products in a topical cream by RP-HPLC. J. Pharm. Biomed. Anal. 2003, 32 (4-5), 921-927.
8. Kokoletsi, M.X; Kafkala, S.; Tsiaganis, M. A novel gradient HPLC method for simultaneous determination of ranitidine, methylparaben and

propylparaben in oral liquid pharmaceutical formulation. *J. Pharm. Biomed. Anal.* 2005, 38 (4), 763-767.

9. Belgaied, J.E; Trabelsi, H. Determination of cisapride, its oxidation product, propyl and butyl parabens in pharmaceutical dosage form by reversed-phase liquid chromatography. *J. Pharm. Biomed. Anal.* 2003, 33 (5), 991-998.

10. Lee, M.R.; Lin, C.Y.; Li, Z.G.; Tsai, T.F. Simultaneous analysis of antioxidants and preservatives in cosmetics by supercritical fluid extraction combined with liquid chromatography-mass spectrometry. *J. Chromatogr. A* 2006, 1120 (1-2), 244-251.

11. Grosa, G.; Grosso, E.D.; Russo, R.; Allegrone, G. Simultaneous, stability indicating, HPLC-DAD determination of guaifenesin and methyl and propyl-parabens in cough syrup. *J. Pharm. Biomed. Anal.* 2006, 41 (3), 798-803.

12. Korsic, J.; Milivojevic, D.; Smerkolj, R.; Kucan, E.; Prosek, M. Quantitative analysis of some preservatives in pharmaceutical formulations by different chromatographic methods. *J. High Resolut. Chromatogr.* 2005, 4 (1), 24-26

13. Radus, T.P.; Gyr, G. Determination of antimicrobial preservatives in pharmaceutical formulations using reverse-phase liquid chromatography. *J. Pharm. Sci.* 1983, 72(3), 221-224.

14. Shabir, G.A.; Lough, W.J.; Shafique, A.A.; Bradshaw, T.K. Evaluation and application of best practice in analytical method validation. *J. Liq. Chromatogr. Relat. Technol.* 2007, 30 (3), 311-33.

15. Shabir, G.A. Validation of HPLC methods for pharmaceutical analysis: Understanding the differences and similarities between validation requirements of the U.S. Food and Drug Administration, the U.S. Pharmacopoeia and the International Conference on Harmonization. *J. Chromatogr A* 2003, 987 (1-2), 57-66.

16. Shabir, G.A. Step-by-step analytical methods and protocol in the quality system compliance industry. *J. Validation. Technol.* 2004, 10 (4), 314-324.

17. G.A. Shabir, Sep. Sci. Technol., 45, 670 (2010)

18. ICH Q2(R1), International Conference on Harmonization, Validation of Analytical Procedures: Text and Methodology, ICH, Geneva, November 2005.

19. Reviewer Guidance: Validation of Chromatographic Methods, Food and Drug Administration (FDA), Centre for Drug Evaluation and Research (CDER), November 1994.

20. USP-33, United States Pharmacopeia, United States Pharmacopieal Convention, Rockville, Maryland, USA, 2010

18 DEVELOPMENT AND VALIDATION OF A HPLC METHOD FOR THE DETERMINATION OF PROCAINAMIDE, N-ACETYLPROCAINAMIDE AND CAFFEINE IN TABLET DOSAGE FORMULATION

ABSTRACT

This paper presents a new reversed-phase high-performance liquid chromatographic method with diode-array UV absorbance detection for the determination of procainamide, N-acetylprocainamide and caffeine drugs. The separation was achieved with acetonitrile-0.05 M KH_2PO_4 (10:90, v/v) as mobile phase, a Hypersil C18 (150 × 4.6 mm) column and spectrophotometric detection was carried out at 254 nm. The linear range of determination for procainamide and N-acetylprocainamide were 40-160 µg/mL and for caffeine 20-80 µg/mL. The relative standard deviation values for intermediate precision studies were <1%. Statistical analysis of the data showed that the method was precise, accurate, reproducible and selective for the analysis of procainamide, N-acetylprocainamide and caffeine drugs. The method was successfully employed for the determination of procainamide in commercially available tablet dosage form.

Keywords: RP-HPLC, Procainamide, N-acetylprocainamide, Caffeine, Tablets, Method validation

1. INTRODUCTION

Procainamide (PA) is an antiarrhythmic drug that has been commonly used for over two decades in the treatment of premature ventricular contractions, ventricular tachycardia and atrial fibrillation. It's also used as secondary to lidocaine in patients who are allergic to lidocaine. Procainamide as the free base has a pK_a of 9.23. Procainamide is used as an active in pharmaceutical preparations such as capsules, tablets and injections in different potencies. N-acetylprocainamide (NAPA) is the major metabolite of procainamide, also has an antiarrhythmic effect with potency similar to that of procainamide [1,2]. The estimation of procainamide and N-acetylprocainamide is a more effective guide to therapy than the determination of procaiamide concentration alone. Caffeine (CF) is a drug that is naturally produced in the leaves and seeds of many plants. Caffeine is also produced artificially and added to certain foods and pharmaceutical preparations such as oral suspension and tablet dosage forms. Caffeine is defined as a drug because it stimulates the central nervous system, causing increased alertness. It is frequently included in compound analgesic preparation with aspirin or codeine. Caffeine is in tea, coffee, chocolate, many soft drinks and pain relievers and used other over-the-counter (OTC) drugs. The chemical structures for PA, NAPA and CF are shown in Figure 1. Some existing analytical procedures are available in literature for the determination of procainamide, N-acetylprocainamide and caffeine, either alone or in combination with other drugs by normal phase high-performance liquid chromatography (HPLC) and other techniques [3-8] in plasma.

These all methods are specific but are slow with longer retention time for NAPA and even poor separation. Furthermore, these reported methods are not meet current validation regulatory requirements. Literature survey revealed that the reversed-phase HPLC method of analysis has not been explored for PA, NAPA and CF in pharmaceutical dosage forms until to date. Therefore it was felt necessary to develop a reversed-phase HPLC method for simultaneous determination of these compounds. In view of this, the present study describes the development of a new, simple, rapid and robust reversed-phase HPLC method with diode-array UV absorbance detection (DAD UV) for the simultaneous determination of PA, NAPA and CF in a single chromatographic run from pharmaceutical dosage forms. As a best practice [9-11] in the subsequent investigation, the new reversed-phase HPLC with diode-array detection assay method was validated according to criteria described in the literature [12-15] and its applicability was evaluated in commercial pharmaceutical tablet dosage forms.

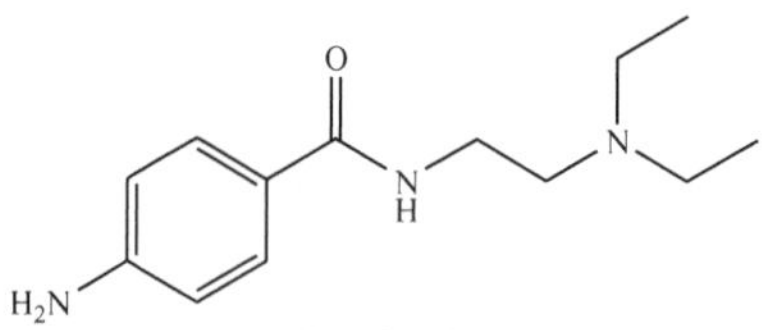

Procainamide

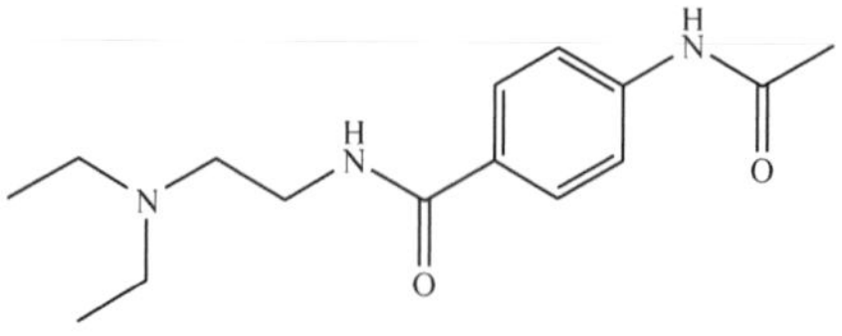

N-Acetylprocainamide

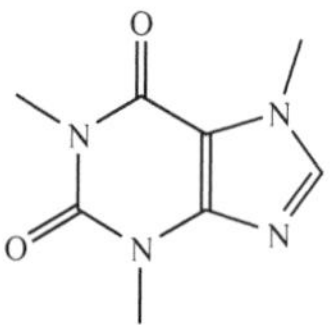

Caffeine

FIGURE 1 Chemical structures of the investigated compounds.

2. EXPERIMENTAL

2.1 Materials and Methods

Acetonitrile (HPLC-grade) was obtained from Fisher Scientific (UK). Procainamide (4-amino-*N*-(2-diethylaminoethyl) benzamide), N-acetylprocainamide (4-acetylamino-*N*-(2-diethylaminoethyl) benzamide), caffeine (1,3,7-trimethyl- 1*H*-purine- 2,6(3*H*,7*H*)-dione) and potassium dihydrogen phosphate (KH$_2$PO$_4$) were purchased from Sigma-Aldrich (UK). Commercial tablets were obtained from local pharmacy (Oxford, UK). Ultra-purified (deionised) water was prepared in-house using a Milli-Q water system (Millipore, UK).

A Knauer HPLC system (Berlin, Germany) equipped with a model 1000 LC pump, an online degasser, model 3950 autosampler, and model 2600 ultraviolet diode-array detector was used. The data were acquired via Knauer ClarityChrom Workstation data acquisition software. A reversed-phase Hypersil C18 (150×4.6 mm, 5 μm) column was used at ambient temperature. The mobile phase consists of acetonitrile-0.05M KH_2PO_4 (10:90, v/v) solution. The mobile phase was filtered using 0.45 μm membrane filter and was continuously degassed on-line. The flow rate was 1 mL/min and the injection volume was 10 μL. The diode-array UV absorbance detection wavelength was 254 nm.

2.2 Standard and Sample Preparation

A combined standard stock solution of accurately weighted PA (10 mg), NAPA (10 mg) and CF (5 mg) were prepared in 100 mL volumetric flask and dissolved in mobile phase, yielding a final concentration of 0.1, 0.1 and 0.05 mg/mL, respectively. The linearity solutions containing PA and NAPA (40-160 μg/mL) and CF (20-80 μg/mL) were prepared in the mobile phase.

The mean weight of finally powdered tablets containing 500 mg of PA was accurately transferred into 200 mL volumetric flask and about 160 ml of mobile phase was added. This mixture was extracted in the ultrasonic bath for 10 min at room temperature and diluted with mobile phase to the mark. The supernatant liquid was filtered through Millipore 0.22 µm filter. One millilitre of this solution was transferred to the 25 mL volumetric flask and diluted with mobile phase to the mark.

3. RESULTS AND DISCUSSION

3.1 Method Development

The procedure for the simultaneous determination of PA, NAPA and CF using HPLC/DAD UV is reported. The mobile phase was chosen after several trials with methanol, acetonitrile, water and buffer solutions in various compositions and at different pH values. The best separation was obtained using the mobile phase consisted of a mixture of acetonitrile and 0.05M KH_2PO_4 in ratio of 10:90 (v/v). Flow rates between 0.5 and 1.5 mL/min were studied. A flow rate of 1 mL/min gave an optimal signal to noise ratio with excellent resolution (>2) and analysis time (<12 min). The diode-array detector was set at 200 to 400 nm and PA, NAPA and CF were extracted at maximum absorption at 254 nm (Figure 2) and this detection wavelength was chosen for the assay method. Preliminary experiments were performed to select the column most suitable for the separation of the analytes consider the physical and chemical properties of the analyte(s), the mode of analysis and how the analytes will interact with the surface of the chromatographic phase. Using Hypersil C18 column, the retention times for PA, NAPA and CF were found to be 2.92, 7.12, and 10.55 min, respectively (Figure 3). For the determination of method robustness, a number of chromatographic parameters were determined, which included flow rate, temperature, mobile phase composition, wavelength accuracy and column from different lots. In all cases, good separations of PA, NAPA and CF were always achieved, indicating that the method remained selective for PA, NAPA and CF under the tested conditions.

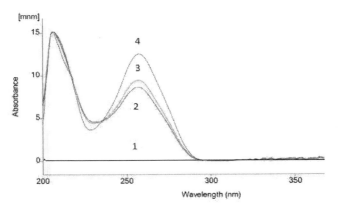

FIGURE 2 DAD-UV spectra of blank (1), procainamide (2), N-acetylprocainamide (3) and caffeine (4).

System suitability testing verifies that the HPLC system is working as expected. It is based on the concept that the equipment, electronics, analytical operations and samples to be analysed constitute an integral system. The system suitability parameters were calculated experimentally using the USP criteria including number of theoretical plates (column efficiency), capacity factor, resolution, tailing factor and relative standard deviation (precision). Theoretical plate number (N) is a measure of column efficiency, that is, how many peaks can be located per unit run-time of the chromatogram and was calculated using formula $N = 16 \, (t_R / t_w)^2 = L$

/H. N is fairly constant for each peak on a chromatogram with a fixed set of operating conditions. H, or the height equivalent of a theoretical plate (HETP), measures the column efficiency per unit length (L) of the column. Capacity factor (k) was calculated using the equation $k' = (t_R - t_m)/t_m$, where tm is unretained peak's retention time and t_R is retention time of the peak of interest. In practice the k value for the first peak of interest should be > 1 to assure that it is separated from the solvent. Tailing factor was calculated using formula $T = W/2f$, where W is width at 5% of the peak height and f is distance between maximum and the leading edge of the peak. The peak resolution was calculated using formula $R_s = 2 (t_{R2} - t_{R1}) / (w_2 + w_1)$, where t_w is peak width measured at baseline of the extrapolated straight sides to baseline. The reproducibility of the HPLC system and method was measured as relative standard deviation (RSD) from peak areas and retention times of six replicate injections.

In the present study, system suitability was performed by injecting six replicate injections of a solution containing 0.1, 0.1 and 0.05 mg/PA, NAPA and CF/mL at the beginning of the validation run. The RSD of the peak area responses was measured, giving an average of 0.22% (n = 6). The tailing factor (T) for PA, NAPA and CF peaks were found to be 0.93, 1.02 and 1.16; the theoretical plate number (N) 3527, 4976, 4326; capacity factor (k') 2.42, 6.34 and 10.16; the resolution (R_s) 11.01 and 6.39 respectively (Table 1). The retention time (t_R) variation RSD for each analyte was <1% for six injections.

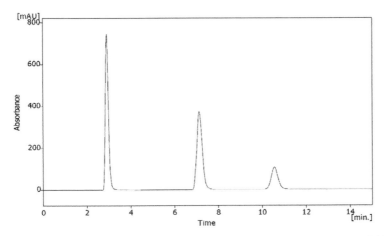

FIGURE 3 HPLC chromatogram of reference standard procainamide (t_R 2.92), N-acetylprocainamide (t_R 7.12) and caffeine (t_R 10.55 min).

The stability of standard solutions was investigated at intervals of 24 and 48 h. The stability of solutions was determined by comparing area% and peak purity results for PA, NAPA and CF. The area% values were within 0.4% after 48 h.

TABLE 1 System suitability test parameter of the HPLC method

Parameters	Acceptance limits	Test results		
		PA	**NAPA**	**CF**
t_R (min)	-	2.92	7.12	10.55
R_s	$R_s > 1.5$	-	11.01	6.39
K'	> 2	2.42	6.34	10.16
T	≤ 2	0.93	1.02	1.16
N	>2000	3527	4976	4326
Injection precision (n = 6)	RSD $\leq$ 1 (%, $n \geq$ 5)	0.17	0.11	0.22

3.2 Method Validation

The linearity test was performed using seven different amounts of PA and NAPA in the range 40-160 µg/mL and for CF 20-80 µg/mL. Solutions corresponding to each concentration level were injected in triplicate and linear regression analysis of the PA, NAPA and CF peak area (y) versus PA, NAPA and CF concentration (x) was calculated. The correlation coefficients ($r^2 \geq 0.9997$) obtained for each drug for the regression line demonstrates that there is a strong linear relationship between peak area and concentration of PA, NAPA and CF (Table 2).

TABLE 2 Linearity assessment of the HPLC method

Analytes	Percent of nominal	Concentration (µg/mL)*	Range (µg/mL)	Equation for regression line	R^2
PA	40-160	100	25-160	y = 51.227x + 110.61	0.9999
NAPA	40-160	100	20-128	y = 49.88x + 1551.3	0.9998
CF	40-160	50	5-40	y = 41.846x + 278.68	0.9997

*Target concentration corresponding to 100%.

Recovery studies may be performed in a variety of ways depending on the composition and properties of the sample matrix. In the present study, three different solutions were prepared with a known added amount of pure PA, NAPA and CF compounds to give a concentration range of 80-120% of that in a test preparation. These solutions were injected in triplicate and percent recoveries of response factor (area/concentration) were calculated (Table 3).

TABLE 3 Accuracy studies of the HPLC method

Analytes	Applied Concentration (% of Target)		
	80	100	120
PA	99.27±0.69*	99.92±0.37	99.43±0.74
NAPA	98.83±0.52	98.68±0.71	98.84±0.27
CF	99.76±0.55	99.27±0.15	99.32±0.58

*Mean ± coefficient of variation (%CV), n = 3

The precision of the method was determined by repeatability (intra-day) and intermediate precision (inter-day variation). Repeatability was examined by analysing ten determinations of the same batch of each component at 100% of the test concentration. The RSD of the areas of PA, NAPA and CF peak were found to be less than 0.18% (Table 4), which confirms that the method is sufficiently precise. Intermediate precision (inter-day variation) was studied by assaying five samples at three concentration levels (80, 100, and 120%) of PA, NAPA and CF on different days. Solutions corresponding to each concentration level were injected in duplicate. The RSD values across the system were calculated and found to be less than 0.29% (Table 4) for each of the multiple sample preparation, which demonstrates excellent precision for the method.

TABLE 4 Method validation results of the HPLC method

Parameter	Components results		
	PA	NAPA	CF
Repeatability (Area %RSD, n = 10)	0.11	0.17	0.13
Intermediate precision (n = 3)			
Day 1, %RSD	0.15	0.09	0.14
Day 2, %RSD	0.24	0.21	0.29
LOD µg/mL (S/N)*	0.50 (S/N = 3.2)	0.52 (S/N = 3.1)	0.16 (S/N = 3.2)
LOQ µg/mL (S/N)	12, (S/N = 10.2)	13.5 (S/N = 10.1)	4.5 (S/N = 10.2)

*Signal-to-noise ratio

Injections of the blank were performed to demonstrate the absence of interference with the elution of the PA, NAPA and CF. This result demonstrated that there was no interference from the other compounds and, therefore, confirm the specificity of the method. Forced degradation studies were also performed to evaluate the specificity of each drug under five stress conditions (heat, UV light, acid, base, oxidative). Solutions of each drug were exposed to 60° for 1h, UV light for 24 h, acid (1M HCl) for 24h, base (1M NaOH) for 4h, and oxidative (3% H_2O_2) for 4 h. A summary data of the stress results is shown in Table 5, which showed no changes in retention times of each drug and no degradation peaks were detected. This was further confirmed by peak purity analysis on a DAD UV detector (Figure 4) and, therefore, confirms the specificity of the method.

TABLE 5 Summary of forced degradation study

Stress conditions	Retention time (min)			Assay (%)*		
	PA	NAPA	CF	PA	NAPA	CF
Reference (fresh solution)	2.92	7.12	10.55	99.94	99.56	99.92
Acid (1 M HCl for 24 h)	2.97	7.15	10.59	99.67	98.26	98.87
Base (1 M NaOH for 4 h)	2.93	7.13	10.57	98.96	98.75	98.17
Heat (60° for 1 h)	2.94	7.14	10.56	99.82	99.17	99.46
light (UV light for 24 h)	2.92	7.12	10.54	99.91	98.97	99.74
Oxidative (3% H_2O_2 for 4 h)	2.94	7.14	10.56	99.84	99.24	99.57

*$n = 5$

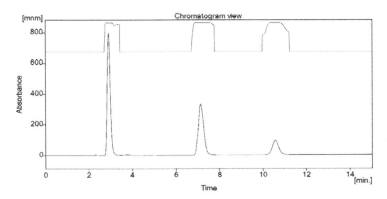

FIGURE 4 HPLC/DAD UV peak purity chromatogram of procainamide (t_R 2.92),n-acetylprocainamide (t_R 7.12) and caffeine (t_R 10.55 min).

The limit of detection (LOD) and limit of quantitation (LOQ) tests for the procedure were performed on samples containing very low concentrations of procainamide, N-acetylprocainamide and caffeine. LOD is defined as the lowest amount of analyte that can be detected above baseline noise, typically, three times the noise level. LOQ is defined as the lowest amount of analyte that can be reproducibly quantitated above the baseline noise that gives a signal-to-noise (S/N) ratio of 10. The LOD was (S/N = 3.2) 0.50, (S/N = 3.1) 0.52 and (S/N = 3.2) 0.16 µg/mL and LOQ was (S/N = 10.2) 12, (S/N = 10.1) 13.5 and (S/N = 10.2) 4.5 µg/mL, and %RSD was < 2% (n = 3) for PA, NAPA and CF, respectively (Table 4).

To demonstrate the applicability of the present method, commercially available three batches of tablets containing 500 mg PA were analysed. Assay results for three samples of tablets expressed as the percentage of the label claim, were found between

99.15-102.04%. Results showed (Table 6) that the content of PA in tablet formulation was to the counter requirements (90-110% of the label claim). The chromatogram obtained from sample analysis is given in Figure 5.

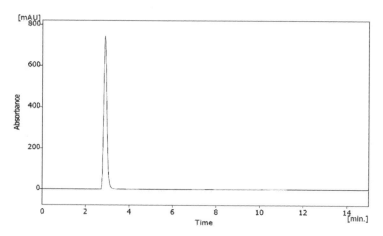

FIGURE 5 HPLC chromatogram obtained from procainamide tablets dosage form. Procainamide peak eluted at 2.92 min.

TABLE 6 Recovery study of procainamide content in tables

Comp. (Lot #)	Label claim (mg)	Found (mg)	n	Recovery (%)	RSD (%)
Procainamide					
1	500	498.76	10	99.15-102.04	0.46
2	500	497.92	10	98.73-101.14	0.72
3	500	498.27	10	99.26-101.27	0.97

4. CONCLUSION

A reversed-phase HPLC/DAD UV method is simple, rapid, sensitive, robust, and therefore suitable for the routine analysis of PA, NAPA and CF in drug substances and final drug products such as tablet dosage forms for marketing release. The proposed method could also be used reliably for stability monitoring of PA, NAPA and CF drugs.

REFERENCES

1. Elson J, Strong JM, Lee WK, Atkinson AJ. Antiarrhythinic efficacy of N-acetylprocainamide in patients with premature ventricular contractions. *Clin. Pharmacot. Ther.* 1975; 17:722-30.

2. Lee WK, Strong JM, Kehoe RF, et al. A ntiarrhythmicefficacy of N-acetylprocainamide in patients with prematureventricular contractions. *Clin. Pharmacol. Ther.* 1976; 19:508- 14.

3. Rocco RM, Abbott DC, Giese RW, Karger BL. Analysis for procainamide and N-acetyl procainamide in plasma or serum by high pressure liquid chromatography. *Clin. Chem.* 1977; 23:705-8.

4. Orville HW, William DM. Rapid determination of procainamide and its *N*-acetyl derivative in human plasma by high-pressure liquid chromatography. *J. Pharm. Sci.* 2006; 66:874-75.

5. Patel CP. Improved liquid chromatographic determination of procainamide and N-acetylprocainamide in serum. *Therapeutic Drug Monitoring* 1983; 5:235-38.

6. Ram NG, Francis E, Diane L. Fluorescence photometric quantitation of procainamide and N-acetylprocainamide in plasma after separation by thin-layer chromatography. *Anal. Chem.* 1978; 50:197-99.

7. Kabra PM, Chen SH, Marton LJ. Liquid chromatographic determination of antidysrhythmic drugs: procainamide, lidocaine, quinidine, disopyramide, and propranolol. Therapeutic Drug Monitoring, 1981; 3:1-113.

8. Dutcher JD, Strong JM. Determination of plasma procainamide and N-acetylprocainamide concentration by high pressure liquid chromatography. *Clin. Chem.* 1977; 23:1318-20.

9. Shabir GA, Lough WJ, Shafique AA, Bradshaw TK. Evaluation and application of best practice in analytical method validation. *J. Liq. Chromatogr. Rel. Technol.* 2007; 30:311-33.

10. Shabir GA, A practical approach to validation of HPLC methods under current good manufacturing practices. *J. Validation Technol.* 2004; 10:210-18.

11. Shabir GA. Systematic strategies in high-performance liquid chromatography method development and validation. *Sep. Sci. Technol.* 2010; 45:670-80.

12. Shabir GA. Validation of HPLC methods for pharmaceutical analysis: Understanding the differences and similarities between validation requirements of the U.S. Food and Drug Administration, the U.S. Pharmacopoeia and the International Conference on Harmonization. *J. Chromatogr. A* 2003; 987:57-66.

13. Shabir GA. Step-by-step analytical methods and protocol in the quality system compliance industry. *J. Validation Technol.* 2004; 10:314-24.

14. ICH, Q2(R1), International Conference on Harmonization, Validation of analytical procedures: Text and methodology, Geneva, Nov. 2005.

15. U.S. Pharmacopeia 33. General Chapter 1225, Validation of compendial methods, United States Pharmacopieal Convention: Rockville, Maryland, 2010.

19 DEVELOPMENT AND VALIDATION OF A HPLC METHOD FOR THE DETERMINATION OF METHAMPHETAMINE AND PROPRANOLOL IN TABLETS

ABSTRACT

A new reversed-phase high-performance liquid chromatographic (HPLC) method with ultraviolet diode-array detection (UV-DAD) has been developed and validated for analysis of methamphetamine and propranolol in tablet formulations. The compounds were separated on an XTerra RP_{18} column with acetonitrile-50 mM pyrrolidine (pH 11.5) 50:50 (v/v) as mobile phase at flow rate of 1 mL/min. The analysis was performed with ultraviolet (UV) detection at 214 nm. The analysis time was < 4 min. The calibration curves showed good linearity over the working concentration range of 0.075-0.6 mg/mL. The correlation coefficients were $r^2 = \geq 0.9998$ in each case. The relative standard deviation values for intra- and inter-day precision studies were <1%. Statistical analysis of the data showed that the method is precise, accurate, reproducible and selective for the analysis of methamphetamine and propranolol drugs. The method was successfully employed for the determination of methamphetamine and propranolol in commercially available propranolol tablet formulations.

Keywords: HPLC/UV-DAD, Validation, Methamphetamine, Propranolol, Tablet formulation

1. INTRODUCTION

Methamphetamine is a popular recreational drug that also has some historical use as a therapeutic agent. The drug increases alertness and vigor, and in high doses, can induce perceptions of mental and emotional pleasure, enhance self-esteem, and increase sexual pleasure [1, 2]. It has a history as a periodically popular drug of abuse, which at the time of writing is undergoing resurgence in popularity [3]. Methamphetamine (Figure 1) is the common name for n,α-dimethylphenethylamine, also referred to as desoxyephedrine, methylamphetamine, phenylisopropylmethylamine, and a variety of other similar systematic names. Methamphetamine is an amphetamine derivative and belongs to the class of amphetamines. Methamphetamine is a prototypical basic drug (pK_a 9.9), and is readily extracted from biological material into organic solvents at alkaline pH. Propranolol is a beta-adrenoceptor blocking drug (beta-blocker) mainly used in the treatment of angina and hypertension. It is the only drug proven effective for the prophylaxis of migraines in children. Propranolol (Figure 1) is available in generic forms in tablet, oral solutions and syrups formulations as propranolol hydrochloride. Propranolol is also used to lower blood pressure, abnormal heart rhythms, heart disease and certain types of tremor. Unfortunately, supply of methamphetamine drug has increased dramatically on the European illegal market [4,5], including Macedonia. Consequently, the analysis of methamphetamine has become of increased interest from a point of view of toxicology, occupational medicine and law enforcement. The literature presently describes only one analytical method for analysis of methamphetamine and its metabolites in plasma using high-performance liquid chromatography (HPLC) with C6 column [6]. Literature survey revealed that the reversed-phase HPLC method of analysis has not been explored for methamphetamine and propranolol in single chromatographic run until to date. Therefore it was felt necessary to develop a HPLC method for determination of methamphetamine and propranolol. In view of this, the present study describes the development of a new, simple, rapid and robust HPLC method with ultraviolet diode-array detection (UV-DAD) for the determination of methamphetamine and propranolol drug substance and from tablet formulations. Analytical method development and validation is an important part of analytical chemistry and plays a major role in the discovery, development, and manufacture of medicinal products. The official test methods that result from these processes are used by quality control laboratories to ensure the identity, purity, potency and performance of drug product 'quality' essential for drug safety and efficacy. Frequently, HPLC is the analytical method of choice in medicinal analysis because of its specificity and sensitivity. Finally, the developed analytical method was validated to assess the validity of research data means determining whether the method used during the study can be trusted to provide a genuine, account of the intervention being evaluated. As a best practice [7,8] in the subsequent investigation, the new HPLC/UV-DAD assay method was validated according to criteria described in the literature [9-15].

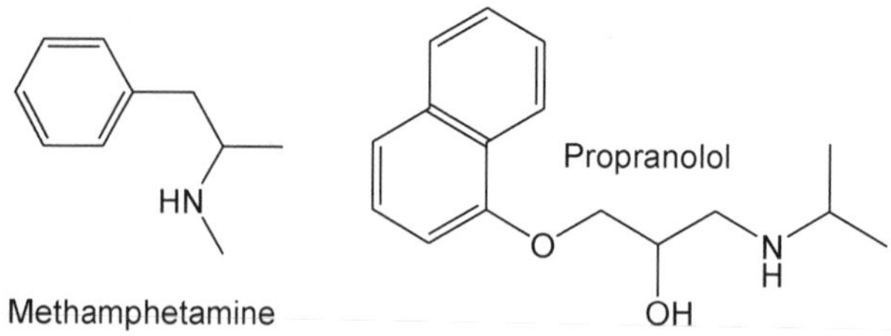

FIGURE 1 Chemical structures of component studied in this work.

2. EXPERIMENTAL

2.1 Materials

All chemicals and solvents were of analytical reagent grades. Acetonitrile was obtained from BDH (Poole, UK). Methamphetamine (n-methyl-1-phenyl-propan-2-amine), propranolol (*RS*)-1-(isopropylamino)-3-(1-naphthyloxy)propan-2-ol) and pyrrolidine (tetrahydropyrrole) were purchased from Sigma-Aldrich (Gillingham, UK). Ultra-purified (deionised) water was prepared in-house using a Millipore Milli-Q water system (Watford, UK). Propranolol tablets were obtained from local pharmacy (Oxford, UK).

2.2 HPLC System and Conditions

Chromatographic separation was carried out on a Knauer HPLC system (Berlin, Germany) equipped with a model 1000 LC pump, an online degasser, model 3950 autosampler, and model 2600 photodiode-array (PDA) detector was used. The data were acquired via Knauer ClarityChrom Workstation data acquisition software. The mobile phase consisted of a mixture of acetonitrile-50 mM pyrrolidine (pH 11.5), 50:50 (v/v) was used, at a flow rate of 1 mL/min. The injection volume was 10 μL and the detection wavelength was set at 214 nm. The run time was 5 min. Reversed-phase LC analysis was performed isocratically at 30°C using XTerra RP_{18} (150 × 4.6 mm, 5 μm particle) column (Waters, UK).

2.3 Standard Preparation

Stock solutions of methamphetamine and propranolol were prepared in mobile phase at concentrations of 1.00 mg/mL (S1). The solutions were stored at room temperature (22°C ±1°C) until analysis. Series of standards for each of the substance was prepared by progressive dilution of the stock solution for calibration study. Fifteen millilitre aliquot of S1 was transferred to another 50 mL volumetric flask and diluted in mobile phase yielding a final concentration of 0.3 mg/mL. Ten millilitre of this assay working solution was injected into the chromatographic system.

2.4 Sample Preparation

The mean weight of finely powdered propranolol tablet containing 80 mg of propranolol was accurately transferred into 50 mL volumetric flask and about 30 mL of mobile phase was added; the mixture was extracted in the ultrasonic bath for 10 min at room temperature and diluted with mobile phase to the mark. The supernatant liquid was filtered through 0.22 μm filter. Two millilitre of this solution was transferred to the 10 mL volumetric flask, diluted with mobile phase to the mark and 10 µL was injected into the chromatographic system.

2.5 Specificity Study

The ability of an analytical method to unequivocally assess the analyte in the presence of other components in the formulation (impurities, degradations, excipients) can be demonstrated by evaluating specificity. The specificity of the method was determined by injecting placebo solution having the same concentration as that of the tablet solution. Forced degradation studies of the tablet sample were also performed. Tablet samples were prepared and degraded under stress conditions such as acidic hydrolysis, basic hydrolysis, oxidative degradation, photo-degradation and thermal degradation for HPLC method. For acid, base and oxidative degradation, samples were individually placed into three volumetric flasks and then 0.1 M HCl, 0.1 M NaOH and 3% H_2O_2 were added separately into the flasks. All the three flasks were then heated in a water bath at 80°C for 4 h. acid and base treated samples were neutralised and all three samples were then diluted to a concentration of 0.3 mg/mL with the mobile phase. For thermal degradation, the sample was exposed to heat at 60°C for 4 h and for photo-degradation, the drug sample was exposed under a UV lamp for 24 h. The samples were withdrawn and analysed using HPLC/UV-DAD.

3. RESULTS AND DISCUSSION

3.1 Development of HPLC Method

The procedure for the simultaneous analysis of methamphetamine and propranolol using isocratic HPLC/UV-DAD method is reported. The mobile phase was chosen after several trials with methanol, acetonitrile, water and buffer solutions in various compositions and at different pH values. The best separation was obtained using the mobile phase consisting of a mixture of 50 mM pyrrolidine (pH 11.5) and acetonitrile (ACN) in ratio of 50:50 (v/v) (Figure 2).

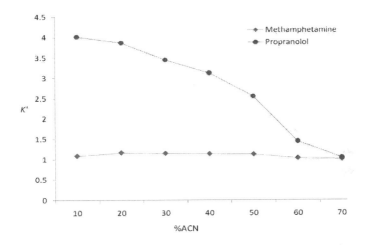

FIGURE 2 Effect of percent acetonitrile (ACN) in the mobile phase on capacity factor (K') for methamphetamine and propranolol.

In RP-HPLC, the pH of the eluent can significantly influence the separation of components. Buffers are required when the sample contains ionic or ionisable analytes. Without a buffer, poor peak shape and variable retention may result. The organic buffer pyrrolidine (pK$_a$ 11.1) was chosen for optimum column life time as phosphate buffers accelerate the dissolution of silica at pH >7. Flow rates between 0.5 and 1.5 mL/min were studied. A flow rate of 1 mL/min gave an optimal signal to noise ratio with excellent separation time. The photodiode-array detector was set at 200 to 400 nm. Methamphetamine and propranolol drug components were extracted at maximum absorption at 214 nm and this wavelength was chosen for the assay method. The separation of basic compounds requires special RP chromatographic sorbents. Retention, selectivity and peak symmetry of basic compounds are strongly been influenced by the silica matrix. Strongly distorted peaks of the basic compounds are often observed when unsuitable RP sorbents are used, due to the interaction of the basic compounds with unreacted SiOH groups on the silica matrix [16]. XTerra RP$_{18}$ is a spherical porous silica carrier, in which the starting silica material is optimized in order to prevent any secondary interactions with basic compounds. The usage of this type of column allows separation of basic compounds such as methamphetamine and propranolol with dissociation constants values (pK$_a$) of 9.9 an d 9, respectively [17,18] without the need of ion pair reagents. Using XTerra RP$_{18}$, the retention times for methamphetamine and propranolol were found to be 2.30 and 2.86 min, respectively. Total time of analysis was < 4 min. Using these optimized conditions, typical chromatogram obtained is illustrated in Figure 3.

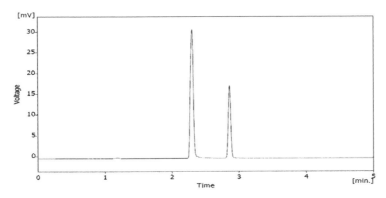

FIGURE 3 HPLC chromatogram obtained from standard methamphetamine (t_R 2.30 min) and propranolol (t_R 2.86 min). LC conditions: mobile phase acetonitrile-50 mM pyrrolidine (pH 11.5), 50:50 (v/v), flow rate 1 mL/min, injection volume 10 μL, detection wavelength at 214 nm, temperature 30°C, column XTerra RP$_{18}$ (150 × 4.6 mm, 5 μm).

Robustness verification studies were also performed in the method development phase. The robustness of the analytical method is defined as the measure of its capacity to remain unaffected by small but deliberate variations in the method parameters and provides an indication of its reliability during normal usage. One way to gauge robustness is to examine some relevant factors, which might influence the reliability of the developed method. Selected factors, namely the mobile phase composition (±2 mL), flow rate (±0.1 unit), temperature (±2°C), wavelength (±5 nm) and column from different lots were investigated. In all cases, good separations of both drug components were always achieved, indicating that the analytical method remained selective and robust under the optimized conditions.

System suitability testing verifies that the HPLC system is working as expected. It is based on the concept that the equipment, electronics, analytical operations and samples to be analysed constitute an integral system. System suitability was evaluated by injecting a solution of methamphetamine and propranolol drugs at 100% test concentration (0.3 mg/mL) in six replicates at the beginning of the validation run. System suitability parameters calculated from the chromatogram (Figure 3), such as peak capacity factor (K'), tailing factor (T), resolution factor (R_s), theoretical plate numbers (N, column efficiency) and percent relative standard deviation (RSD) of peak areas are given in Table 1. The obtained values of these parameters ($1 < k' < 10$, $T \leq 2.0$, $R_S > 2$, $N > 2000$) show that the proposed chromatographic conditions are suitable for separation of the analysed drug components. The values for the injection repeatability (RSD% < 2.0, $n = 6$) show that the system is precise.

TABLE 1 System suitability parameters for the HPLC method

Parameter	Methamphetamine	Propranolol
t_R (min)	2.30	2.86
k'	1.14	2.56
R_s	-	5.23
T	1.13	1.14
N	4362	5026
Repeatability (%RSD, $n = 6$)	0.36	0.29

The stability of analyte stock solutions with mobile phase as solvent, stored at room temperature, was studied for 48 h. No analyte (methamphetamine or propranolol) degradation was detected during this time period (< 2%).

3.2 Validation of the Method

3.2.1 Linearity and range

Linearity test solutions were prepared by diluting stock solution (1 mg/mL) at six concentration levels from 25 to 200% of analytes concentration (0.075-0.6 mg/mL) for both drugs. The solution was injected in triplicate and curves were obtained by plotting the peak area against concentration of the drugs. Linear calibration curves were generated using least-squares linear regression analysis. The mean of two different calibration graphs yielded the following equations: $y = 203.17x - 0.2552$ ($R^2 = 0.9998$) for methamphetamine and $y = 133.99x + 5.0261$ ($R^2 = 0.9999$) for propranolol (Table 2). An excellent correlation exists between the peak areas and concentration of methamphetamine and propranolol.

3.2.2 Precision

The precision of the test method was demonstrated (%RSD) by intera-day and inter-day variation studies. The intera-day studies were carried out by injecting six repeated injections of standard solution of 0.3 mg/mL on the same day, by one analyst under the same experimental conditions. The %RSD values for peak areas were found to be 0.27% and 0.19%, respectively (Table 2).

TABLE 2 Method validation results of the HPLC method

Parameter	Drug substances	
	Methamphetamine	**Propranolol**
Linearity (0.075-0.6 mg/mL)		
Equations (r^2)	y = 203.17x - 0.2552 (0.9998)	y = 133.99x + 5.0261 (0.9999)
Intra-day precision (%RSD)		
0.3 mg/mL ($n = 6$)	0.27	0.19
Accuracy / recovery (%)		
0.15 mg/mL (50%)[a]	99.84 ± 0.22[b]	99.91 ± 0.34
0.30 mgmL (100%)	99.98 ± 0.11	99.82 ± 0.28
0.45 mg/mL (150%)	99.62 ± 0.17	101.07 ± 0.21
LOD µg/mL	0.85	0.95
LOQ µg/mL	2	2.5

[a] applied concentration % of target
[b] ($n = 3$) the coefficient of variation (%CV)

The intermediate precision (within-laboratory variation) was studied by two analysts over two consecutive days at three different concentration levels 0.15, 0.3 and 0.45 mg/mL for methamphetamine and propranolol that cover the assay range (80-120%). Three replicate injections were injected for each solution. The mean and %RSD across the two analysts were calculated from the individual relative % purity mean values at the 50%, 100% and 150% testing amounts. The %RSD values for peak areas obtained by both analysts were ≤ 1.0% and met the intermediate precision criteria (RSD < 2.0%) which illustrated the good precision of this analytical method.

3.2.3 Accuracy

The accuracy of an analytical method is determined by how close the test results obtained by that method come to the true value. It can be determined by application of the analytical procedure to an analyte of known purity (for the drug substance) or by recovery studies, where a known amount of standard is spiked in the placebo (for drug product). In the present study, a number of different solutions were prepared with a known added amount of 50% 100% and 150% for methamphetamine and propranolol drug components and injected in triplicate (n = 3). Percent recoveries of response factor (area and concentration) were calculated ranged from 99.62-101.07 % (Table 2) which indicated the accuracy of the method was accurate within the desired range.

3.2.4 Specificity

Injections of the placebo were performed to demonstrate the absence of interference with the elution of the propranolol drug. These results (Figure 4b) demonstrate that there was no interference from the other compounds and, therefore, confirm the specificity of the method. For the further evaluation of the selectivity of the LC method , the forcibly degraded tablet sample solutions prepared by subjecting the tablet samples to such stress conditions as acid, base, heat, light and oxidative agent were determined under the proposed optimised LC conditions. A summary data of the stress results is shown in Table 3, which showed no changes in retention times of drug components and no degradation peaks were observed. This was further confirmed by peak purity analysis on a HPLC/UV-DAD and, therefore, confirms the specificity of the method.

TABLE 3 Results from evaluation of the forced degradation study for propranolol, $n = 3$

Stress conditions	t_R (min)	Assay (%)	RSD (%)
Reference (fresh solution)	2.86*	99.98	0.14
Acidic (0.1 M HCl at 80°c for 4 h)	2.85	99.17	0.32
Basic (0.1 M NaOH at 80°c for 4 h)	2.86	99.67	0.56
Oxidative (3% H_2O_2 at 80°c for 4 h)	2.87	98.77	1.34
Heat (60 °C for 4 h)	2.83	99.89	0.92
Light (UV light for 24 h)	2.85	99.74	0.17

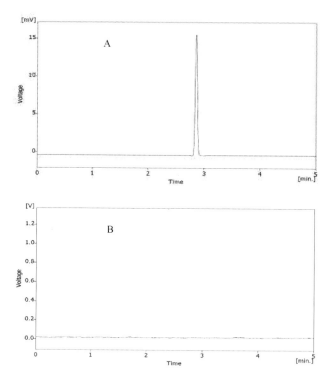

FIGURE 4 HPLC chromatogram obtained from: (a) sample tablet formulation containing 80 mg propranolol (t_R 2.86 min); (b) placebo containing all excipients without active drug. For HPLC conditions, see Figure 3.

3.2.5 Limits of detection and quantitation

The limit of detection (LOD) and limit of quantitation (LOQ) were determined by the calibration plot method. A specific calibration plot was constructed using samples containing amounts of analytes in the range of LOD and LOQ. The values of LOD and LOQ were 0.85 and 0.95 µg/mL and 2 and 2.5 µg/mL for methamphetamine and propranolol, respectively, for 10 µL injection volume. LOD and LOQ were calculated by using the equations: LOD = $Cd \times Syx/b$ and LOQ = $Cq \times Syx/b$, where Cd and Cq are the coefficients for LOD and LOQ, Syx is the residual variance of the regression, and b is the slope. Calculations were performed by using values of Cd and Cq of 3.3 and 10. Precision at the LOQ was checked by analysis of six test solutions prepared at three levels. The %RSD values for peak area was < 5% for LOQ solutions (Table 2), which indicates the sensitivity of the method is adequate.

4. APPLICATION OF THE METHOD

To demonstrate the applicability of the present method, commercially available three batches of tablets containing 80 mg propranolol were analysed. Assay results for three samples of tablets expressed as the percentage of the label claim, were found between 98.66 to 102.11%. Results showed (Table 4) that the content of propranolol in tablet formulation was to the counter requirements (90-110% of the label claim). The chromatogram obtained from sample analysis is given in Fig. 4a. The above results demonstrated that the developed method achieved rapid and accurate determination of compound studied and can be used for the simultaneous determination of methamphetamine and propranolol in drug substances and tablet formulations.

TABLE 4 The determination of propranolol content in tablet formulation

Comp. (Lot No.)	Added (mg)	Found (mg)	n	Recovery (%)	RSD (%)
Propranolol					
1	80	78.93	5	98.66-101.07	1.13
2	80	79.54	5	99.42-102.04	0.86
3	80	79.48	5	99.35-102.11	1.24

5. CONCLUSION

A new reversed-phase HPLC/UV-DAD method has been developed for the determination of methamphetamine and propranolol. Low cost, environment friendly, faster speed analysis, and satisfactory precision and accuracy are the main features of this method. The method was critically validated and statistical analysis of generated high quality data proves that the method is sensitive, specific and robust. The method is stability-indicating and can be conveniently applied for the testing of studied components raw materials, in tablet formulations and batch release by industry.

REFERENCES

1. Mack, A. H.; Frances, R.J.; Miller, S.I. Clinical textbook of addictive disorders, 3rd Ed., the Guilford press, New York, 2005, p. 207.
2. Logan, B.K. Methamphetamine – Effects on human performance and behaviour. *Forensic Sci. Rev.* 2002, 14, 133-151.
3. Anglin, M.D.; Burke, C.; Perrochet, B.; Stamper, E.; Dawud-Noursi, S. History of the methamphetamine problem. *J. Psychoact. Drugs*, 2000, 32, 37.

4. European Union Situation Report on Drug Production and Drug Trafficking 2000-01, EUROPOL, Hague, 2001.

5. Pelegrini, M.; Rosati, F.; Pacifici, R.; Zuccaro, P.; Romalo, F.S.; Lopez, A. Rapid Screening Method for Determination of Ecstasy and Amphetamines in Urine Samples Using Gas Chromatography–Chemical Ionization Mass Spectrometry. *J. Chromatogr. B* 2002, 769, 243-251.

6. Li, N.Y.; Li, Y.; Sellers, E.M. An improved HPLC method for analysis of methamphetamine and its metabolites in plasma. Eur. *J. Drug Metab. Pharmacokinet* 1997, 22, 427-32.

7. Shabir, G.A.; Lough, W.J.; Shafique, A.A.; Bradshaw. T.K. Evaluation and application of best practice in analytical method validation. *J. Liq. Chromatogr. Relat. Technol.* 2007, 30, 311-333.

8. Shabir, G.A. A practical approach to validation of HPLC methods under current good manufacturing practices. *J. Validation Technol.* 2004, 10(3), 210-218.

9. Shabir, G.A. Validation of HPLC methods for pharmaceutical analysis: Understanding the differences and similarities between validation requirements of the U.S. Food and Drug Administration, the U.S. Pharmacopoeia and the International Conference on Harmonization. *J. Chromatogr. A* 2003, 987, 57-66.

10. Shabir, G.A. Step-by-step analytical methods and protocol in the quality system compliance industry. *J. Validation. Technol.* 2004, 10, 314-324.

11. Shabir, G.A. Systematic strategies in high-performance liquid chromatography method development and validation. *Sep. Sci. Technol.* 2010, 45, 670-680.

12. ICH Q2 (R1), International Conference on Harmonization, Validation of Analytical Procedures: Text and Methodology, Geneva, November 2005.

13. Reviewer Guidance: Validation of Chromatographic Methods, Food and Drug Administration (FDA), Centre for Drug Evaluation and Research (CDER), November 1994.

14. Shabir, G.A. HPLC method development and validation for pharmaceutical analysis. *Pharm. Technol. Eur.* 2004, 16(3), 37-49.

15. USP-33, United States Pharmacopeia, General Chapter: 1225, validation of compendia procedures, United States Pharmacopieal Convention, Rockville, Maryland, USA, 2010.

16. Skoog, D.A.; Holler, F.J.; Nieman, T.A. Principles of Instrumental Analysis, Harcourt Brace & Co., Orlando, 1998, p. 739.

17. Laboratory and Scientific Section, United Nations Office on Drugs and Crime Vienna, Recommended Methods for the Identification and Analysis of Amphetamine, Methamphetamine and Their Ring-substituted Analogues in Seized Materials, United Nations, New York, 2006.

18. Noever, R.; Cronise, J.; Relwani, J.R.A. Using Spider-web Patterns to Determine Toxicity. NASA Tech. Briefs. 1995, 19, 82.

20 METHOD DEVELOPMENT AND VALIDATION FOR THE ANALYSIS OF HYDROLYSED GELATINE USING HPLC

ABSTRACT

This paper describes the development and validation of reversed-phase high-performance liquid chromatography (HPLC) method for the assay of hydrolysed gelatine (Gelita-Sol P, trade name). Key chromatographic parameters were investigated including short and long alkyl chains of stationary phases (C_4 and C_{18}), column temperatures (30-60°C) and additives of ion-pairing reagents (trifluoroacetic acid and heptafluorobutyric acid) in the mobile phase. Analytical validation parameters such as specificity and selectivity, linearity, accuracy, precision, limit of detection, limit of quantitation, robustness and system suitability, were evaluated. The calibration curve for hydrolysed gelatine was linear ($r^2 = 0.997$) from 20-200% range of the analytical concentration of 50 mg/mL. The precision of this method was calculated as the relative standard deviation (RSD) was 1.22% ($n = 6$). The RSD for intermediate precision study was 1.77 and recovery of the hydrolysed gelatine ranged between 97.08 and 97.76%. The limits of detection and quantitation were determined to be 5.0 and 10.0 mg/mL, respectively.

Keywords: Hydrolysed gelatine (Gelita-Sol-P), HPLC, Method development, Method validation

1. INTRODUCTION

Gelatine is a collagen derivative, which has a large application in the pharmaceutical, food and adhesive industries as well as photography. In the last decade, many research efforts have been done to develop techniques and methods for the separation, purification and characterisation of hydrolysed gelatine (also called Gelita-Sol P, trade name). Hydrolysed gelatine is a highly purified collagen hydrolysate manufactured from hide. The average molecular weight (MW) is approximately 3k. Hydrolysed gelatine is contains approximately 97% protein content on dry substance. Gelatine is a high molecular weight polypeptide derived from collagen, the primary protein component of animal connective tissues. Industrial preparation of gelatine involves the controlled hydrolysis of the organized structure of collagen to obtain soluble gelatine. The most important sources of collagen for gelatine production are bovine hide, bone and pigskin. Gelatines from different sources can be very similar in their physiochemical properties, which makes their differentiation very difficult. Collagen hydrolysate is manufactured from animal bones and hides. The material is homogenised and washed, and the bones are demineralised with dilute mineral acid. The resulting product, ossein, is practically pure collagen. After alkaline or acid processing, depending on whether the source is bovine or pigskin, respectively, the raw materials are extracted in several stages with warm water.

During this process, the gelatine goes into solution. After concentration, gelatine takes place during the cooling process. Advanced variants of gelatine in the form of gelatine hydrolysate do not gel any further, giving it the advantage of being soluble in cold water.

Pharmaceutical grade collagen hydrolysate (PCH) is a soluble powder obtained by hydrolysis of pharmaceutical gelatine (USP XXII/NF XVIII) by use of enzymatic process with an U.S. Food and Drug Administration (FDA) approved enzyme. There is a final sterilising step before drying. The average molecular weight of PCH ranges from 2,000 to 6,000 Daltons (2 to 6 kD). Its molecular weight is less than the molecular weight of gelatine yet more than the average molecular weight of peptones. Unlike gelatine, PCH does not bind significant amounts of water, but it is disbursable and emulsion stabilizing.

Collagen hydrolysate generally has been regarded as having a low biologic value. It does not contain all of the essential amino acids; tryptophan is not present, and cysteine only in small amounts. However, the protein value of gelatine may relate not only to its amino acid composition, but also to its combined effect with other nutritional proteins. In animal experiments, high-value protein (casein with addition of methionine) can be replaced up to one third by gelatine without animal growth being significantly affected. It is also regarded as a valuable nutritional component because of its excellent digestibility.

The clinical studies have suggested a role for collagen hydrolysate in the management of osteoarthritis based on the postulate that hydrolysed collagen with its abundant amino acids plays a role in cartilage matrix synthesis [1-2]. Gelatine products, which have been used as foods for a number of centuries, are attractive with respect to safety and overall lack of toxicity [3]. Hydrolysed gelatine products have long been used in pharmaceuticals and foods in the United States and Europe.

Research on structural and physico-chemical properties of proteins has been essential for elucidating their molecular structure responsible for their functionality in food and or pharmaceuticals. In addition, the development of methods for the purification of the proteins has been of utmost interest in biotechnology research. In fact, the purity of a protein is a pre-requisite for its structure studies or its application, low degree of purity being requested for industrial application in food and pharmaceutical industries.

If HPLC in different modes is a well-established technique in food and pharmaceuticals protein research, the new emergent technique capillary electro chromatography is expected to have great potential in the separation of proteins [4]. Analytical approaches based on the use of mass spectrometry (MS) are also well established in protein products analysis. A number of review papers on the application of chromatography and MS to proteins have appeared in the literature, attesting a large increase in related publications and the increasing interest and efforts made in this direction [5-10]. To my knowledge, this is the first report on HPLC based separations of this hydrolysed gelatine compound. Herein, my effort in developing HPLC assay method for this important compound is described. The purpose of this study is to develop HPLC separation method for hydrolysed gelatine that can be applied to differentiate between good and bad batches of hydrolysed gelatine materials. In most cases HPLC method development is carried out with ultraviolet (UV) detection using either a variable-wavelength (spectrophotometric) or a diode-array detector (DAD). Therefore, we selected UV detection, which can provide an adequate response for most samples. Alternative detectors can be selected primarily when: Samples have little or no UV absorbance; Analyte concentrations are too low for UV detection; Sample interference is important; Qualitative structural information is required.

2. EXPERIMENTAL

2.1 Materials

Acetonitrile (HPLC grade) was purchased from Merck (Darmstadt, Germany). Trifluoroacetic acid (99.8%, spectrophotometric grade) was obtained from Sigma-Aldrich (St. Louis, MO, USA). Hydrolysed gelatine was supplied by Gelita Europe (Eberbach, Germany).

2.2 Instrumentation

The analytical separations were carried out on a Perkin Elmer (Norwalk, CT) HPLC system, equipped with a model LC 200 UV/Vis detector, series 200 LC pump, series 200 autosampler and series

200 peltier LC column oven. The analytical column was a Jupiter C_4 (250 x 4.6 mm) 5 μm, 300 pore size (Phenomenex, Torrance, CA, USA). The mobile phase consisted of 0.02% (v/v) trifluoroacetic acid (TFA) in filtered deionised water as solvent A and acetonitrile containing 0.02% TFA (v/v) as solvent B. The mixture was pumped as a gradient starting at 98% A and 2% B and was maintained for 2 min. Over a 2 min period the mixture changed to 85% A and 15% B which was maintained for 5 min and then changed to 40% A and 60% B which was maintained for 33 min. The system was equilibrated for 15 min for the next injection. The mobile phase was filtered through 0.45 μm membrane filter and continuously degassed with on-line degasser. The flow rate was 1.20 mL/min. Column temperature was maintained at 35°C. UV detection was measured at 230 nm and the volume of sample injected was 10 μL. The control of the HPLC system and data collection was by a Dell Pentium III computer, linking with 600 interface and equipped with Perkin Elmer Totalchrom software.

2.3 Sample Preparation

All sample solutions at 50.0 mg/mL were prepared by dissolving hydrolysed gelatine in deionised water.

3. RESULTS AND DISCUSSION

3.1 Method Development

Some chromatographic parameters such as column type, mobile phase, and conditioning time were investigated to obtain a good separation of the hydrolysed gelatine analyte within acceptable time span. In RP-HPLC method development, the important parameters for choosing a column include the type of bonded phase, column dimensions, particle size, carbon load, and the degree of end capping. For protein analysis, the scope of RP-HPLC method development has been limited to wide pore, silica-based columns of shorter alkyl chain length and pores perfusion resins of highly cross-linked polystyrene-divinylbenzene to minimize recovery losses of hydrophobic species. RP-HPLC applications for large hydrophobic proteins typically employ n-butyl (C_4) silica-based columns of 5 μm particle size, $\geq$ 300 pore size to obtain adequate loading capacity, recovery, backpressure, and flow rate [11].

Preliminary experiments were performed to select the column most suitable for our purpose: the separation of the hydrolysed gelatine. Three C_{18} and one C_4 column were tried in the following order: Biosuite C_{18}, Hypersil C_{18}, ODS, Hyperclone C_{18} and Jupiter C_4. The different columns showed different selectivity due to the different degree of silanization and different carbon percentage. Hypersil and Hyperclone, did not give a good separation, even when changing the composition of the mobile phase. Biosuite and Jupiter gave satisfactory results in terms of separation, but the final choice was for Jupiter C_4 column because, under the same experimental conditions (see chromatography section), the retention times were shorter and the peaks were sharper than those obtained on the Biosuite C_{18} column. Under these conditions all hydrolysed gelatine peaks are eluted in less than 14 min with acceptable separation.

Mobile phase: ion-pairing reagents are often used in RP-HPLC analysis to shield the effective charge of functional groups on a protein. Within the pH range for chromatography on silica-based columns, the ionizable functional groups include carboxylates (pK_a of 2 and 4), sulfhydryls (pK_a of 8), amines (pK_a of 6,9 and 10, or 11), and guanidines (pk_a of 12 and 13). Two approaches to neutralizing these functional groups are to lower the pH to about 2 to reduce the carboxylate charge and to use anionic ion-pairing reagents to neutralize the positive charged functional group, thereby increasing the hydrophobic nature of the separation. Although standard HPLC methods for protein characterization almost exclusively employ mobile phases containing a default level of 0.1% TFA with a gradient of acetonitrile, the resulting TFA suppression of the mass spectrometric signal makes identification and characterization of low levels of proteins unfeasible. Lower amounts of TFA can be used for good chromatographic peak resolution. Other ion-pairing reagents used less frequently include heptafluorobutyric acid (HFBA) and pentafluoropropionic acid (PFPA).

Initially, we tried both TFA and HFBA of ion-pairing reagents under the same experimental conditions and found good separation with the TFA reagent (Figure 1). The mobile phase consisted of 0.02% TFA (v/v) in filtered deionised water as solvent A and acetonitrile containing 0.02% TFA (v/v) as solvent B (see chromatography section). Peak 1, 4 and 5 are very reproducible in every commercial hydrolysed gelatine batch tested in the present study (see applications of the method section). Peak 5 is the principal hydrolysed gelatine. The differences between good and bad hydrolysed gelatine batches were seen in the region of peak 2 and 3. Some batches gave more than 2 peaks in this region.

The optimal wavelength for hydrolysed gelatine detection was established using two UV absorbance scans over the range of 190 to 400 nm, one scan of the mobile phase and the second of the analytes in the mobile phase. It was shown that 230 nm is the optimal wavelength to maximize the signal. Method development work demonstrated that column temperature had a major impact on separation and recovery of proteins. A sample of hydrolysed gelatine 50 mg/mL was run on HPLC using column temperatures ranging from 30 to 60°C (Figure 2). The recovery and peak shape was optimised at approximately 35°C, while at lower temperatures excessive peak tailing, poor recovery, and column fouling were observed. With an optimised column temperature of 35°C, good separation of the main component was achieved using Jupiter C_4 column of 5 μm particle size and 300 pore size described in section chromatography.

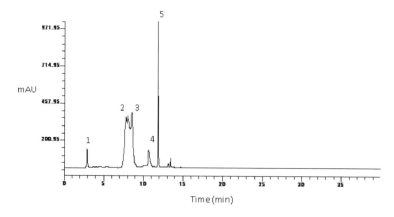

FIGURE 1 HPLC chromatogram of hydrolysed gelatine.

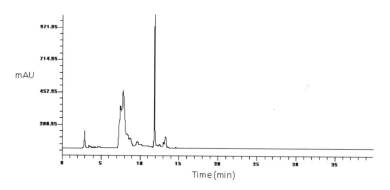

FIGURE 2 HPLC chromatogram of hydrolysed gelatine at 60°C.

To evaluate the quantitative nature of the method, a series of samples were run to test the linearity, range and recovery. Using a Jupiter C_4 column, linearity was assessed by injecting eight reference standards that ranged in concentration from 5 to 200 mg/mL. The integrated peak areas were plotted versus amount injected. The calibration curve was found to be linear from concentration range10-100 mg/mL with a correlation coefficient of 0.996. On the bases of these data, the best concentration was chosen as a working concentration for the assay. The linearity study also showed as hydrolysed gelatine concentration increase the column performance decreased due to sample/column overloaded. At high concentration of compound hydrolysed gelatine in the mobile phase (> 100 in Figure 3) linear isotherm behavior is no longer observed, with predictable effects on the separation. Additional analysis was performed to assess the injection reproducibility and robustness of the chromatography.

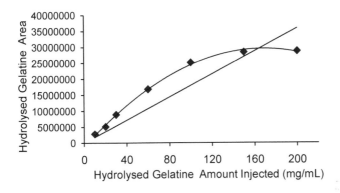

FIGURE 3 Typical calibration graph obtained after analysis of hydrolysed gelatine to demonstrate column overloaded with high hydrolysed gelatine concentration.

A system suitability test was developed for the routine application of the assay method. Prior to each analysis, the chromatographic system must satisfy suitability test requirements (resolution and repeatability). Peak-to-peak resolution, between each peak measured on a reference solution, must be above 1.0. The percent relative standard deviation (%RSD) of the response factor (area:-mass ratio) for hydrolysed gelatine sample peaks was determined from seven replicate injections of the reference solutions and is required to be less than 2.0%.

System suitability testing was performed to determine the accuracy and precision of the system by making seven injections of a solution containing 50 mg of hydrolysed gelatine/ml. All peaks were well resolved, and the precision of injections for all peaks was acceptable. The percent relative standard deviation (%RSD) of the peak area response was measured. The %RSD of peak areas averaged 1.92% (n = 7); the tailing factor (T) for each peak of the hydrolysed gelatine was 1.08, theoretical plate number (N) was 101624.98 and resolution was >5.14 for the main hydrolysed gelatine peak. The retention time variation %RSD was 0.20 for seven injections.

Selectivity was also studied over extended time using several columns and many different batches of mobile phase. Relative RT ranges (RT of peak of interest/RT of hydrolysed gelatine) were as follows: peak 1=7.72; peak 2=8.66; peak 3=9.75; peak 4=10.92; peak 5=11.88. This data indicate that the RT windows for each impurity/amino acid are unique and do not overlap. Overall selectivity was established through determination of purity for each peak using the PDA UV detector.

For assessment of method robustness within a laboratory, a number of chromatographic parameters were varied, which included flow rate, temperature, mobile phase composition, and columns from different lots. In all cases, good separations were always achieved, indicating that the method remained selective for hydrolysed gelatine components under the tested conditions.

The stability study of hydrolysed gelatine solutions was also investigated. The solutions were stable during the investigated eight days and the %RSD was in between 0.06 and 0.10% for retention time for the main hydrolysed gelatine peak 5 at 11.87 min. Standard solutions stored in a capped volumetric flask on a laboratory bench under normal lighting conditions for eight days, were shown to be stable with no significant change in hydrolysed gelatine concentration over this period. Based on these data that show quantitative recovery through eight days, solutions of hydrolysed gelatine and be assayed within eight days of preparation.

3.2 Validation of the Method

Validating analytical methods is a crucial component of successful product development, testing and quality. All product types require some level of evaluation and testing either at the raw material, intermediates, or final product level. Critical decisions may be made based on these results, making it imperative that pharmaceutical and diagnostic companies ensure their accuracy and reproducibility to remain compliant with regulatory guidelines [12-13] in the current climate of increased enforcement. The step-by-step written and approved protocol for test method validation should be followed [14].

3.2.1 Linearity

The linearity of the method should be tested in order to demonstrate a proportional relationship of response versus analyte concentration over the working range. It is usual practice to perform linearity experiments over a wide range of analyte. This gives confidence that the response and concentration are proportional and consequently ensures that calculations can be performed using a single reference standard, rather than the equation of a calibration line. In this study, linearity was studied using five solutions in the concentration range 10-100 mg/mL ($n = 3$). The regression equation was found by plotting the peak area (y) versus the hydrolysed gelatine concentration (x) expressed in mg/mL. The correlation coefficient ($r^2 = 0.997$) obtained for the regression line demonstrates that there is a strong linear relationship between peak area and concentration of hydrolysed gelatine (Table 1).

TABLE 1 Linearity study of the HPLC method for the assay of hydrolysed gelatine.

Concentration of hydrolysed gelatine (mg/mL)	Concentration as percent of 50 mg/mL of hydrolysed gelatine	Hydrolysed gelatine peak area as mean of two injections (μV s)
10	20	2795601
20	40	5157972
30	60	8918107
60	120	16837051
100	200	25162025

Correlation coefficient: 0.997; Intercept (%): -73; Equation for regression line: $y = 250897x + 734692$

3.2.2 Precision (repeatability and intermediate precision)

The precision of the chromatographic method, reported as %RSD, was estimated by measuring repeatability (intra-day assay precision) on six replicate injections at 100% test concentration (50 mg/mL) and intermediate precision (inter-day variation) was studied for two days using three solutions in the concentration range 30, 50, 80 mg/mL ($n = 3$). The %RSD values for the t_R (min) and peak area were found to be less than 2.0% in all cases and illustrated good precision for the analytical method.

3.2.3 Accuracy / recovery studies

The accuracy of an analytical method is determined by how close the test results obtained by that method come to the true value. It can be determined by application of the analytical procedure to an analyte of known purity (for the drug substance) or by recovery studies, where a known amount of standard is spiked in the placebo (for the drug product). In this study, a number of different solutions were prepared with a known added amount of hydrolysed gelatine and injected in triplicate. Percent recoveries of response factor (area and concentration) were calculated as shown in Table 2, and it is evident that the method is accurate within the desired range.

TABLE 2 Recovery studies.

Concentration range (mg/mL)	Recovery (%) (n =3)	RSD (%)
10	97.76	1.00
30	97.74	1.37
50	97.15	1.47
80	97.08	1.33

3.2.4 Specificity and selectivity

Forced degradation studies were performed to evaluate the specificity of hydrolysed gelatine and its impurities under four stress conditions (heat, UV light, acid, base). A summary of the stress results is shown in Table 3 and chromatograms in Figure 4. It is evident from Figure 1 that the method has been able to separate the peaks due to the degraded products from that of the hydrolysed gelatine. This was further confirmed by peak purity analysis on a PDA UV detector.

TABLE 3 Assay (%) of hydrolysed gelatine under stress conditions.

Stress conditions	Sample treatment	t_R (min)	Assay (%)	Peak area (μV s)
Reference	Fresh solution	11.85	98.13	11365875
Acid	1 N HCL for 24 h	11.83	96.53	8819136
Base	1 N NaOH for 4 h	11.91	94.37	9952468
Heat	60 °C for 1 h	11.87	98.25	11294417
Light	UV Light for 24 h	11.85	99.04	10725275

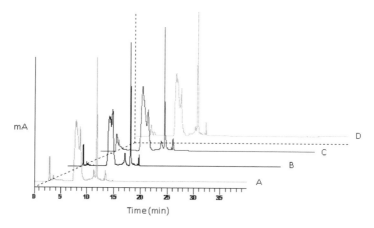

FIGURE 4 HPLC chromatograms of hydrolysed gelatine under stress conditions (A) UV light; (B) heat at 60 °C; (C) base; (D) acid.

3.2.5 Limit of detection and quantitation

The limit of detection (LOD) and limit of quantitation (LOQ) tests for the procedure were performed on samples containing very low concentrations of analyte. LOD is defined as the lowest concentration of analyte that can be detected above baseline noise. Typically, this is three times the noise level. LOQ is defined as the lowest concentration of analyte that can be reproducibly quantitated above the baseline noise with a signal to noise ratio of 10. In this study, the LOD was 5 mg/mL and the LOQ was 10 mg/mL and %RSD 1.89% (n = 3).

4. APPLICATION OF THE METHOD

The developed method was applied to the assay of nine different commercial batches of hydrolysed gelatine raw material. Different peaks were eluted in different good and bad hydrolysed gelatine batches (Figure 5). Table 4 shows the retention times, the area percentages and the precision of each batch calculated from the principal hydrolysed gelatine peak 5.

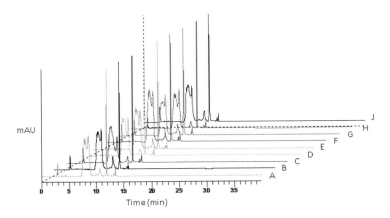

FIGURE 5 HPLC chromatograms of different nine hydrolysed gelatine batches.

TABLE 4 Assay (%) from nine different commercial hydrolysed gelatine batches

Batch #	t_R (min) ± RSD (%)	Area (%) ± RSD (%) ($n = 2$)
A	11.85 ± 0.35	97.27 ± 0.02
B	11.86 ± 0.00	98.13 ± 0.25
C	11.88 ± 0.05	96.70 ± 0.03
D	11.86 ± 0.35	98.72 ± 0.05
E	11.85 ± 0.00	96.09 ± 0.02
F	11.86 ± 0.17	98.88 ± 0.11
G	11.88 ± 0.35	98.09 ± 0.06
H	11.87 ± 0.17	98.75 ± 0.01
J	11.86 ± 0.11	96.11 ± 0.04

5. CONCLUSION

A reversed-phase HPLC method for the separation of complex hydrolysed gelatine is developed and validated that can be reliably applied to differentiate between good and bad hydrolysed gelatine batches. The developed method was applied to nine different commercial hydrolysed gelatine batches and results showed significant differences between good and bad batches. The validation study shows good linearity, sensitivity, accuracy and precision. The suggested technique can be used in quality control for release for hydrolysed gelatine materials.

REFERENCES

1. Oberschelp U. Individual arthrosis therapy is possible. Therapiewoche. 1985: 35, 5094-5097.
2. Seeligmuller K, Happel HK, Can a mixture of gelatine and l-cystine stimulate proteoglycan sysnthesis? Therapiewoche. 1989: 39, 3153-3157.

3. Brook MJV, Lyster SC, Graham BE. Intravenously administered gelatine: a toxicity study. *J Lab Clin Med*. 1947: 32(9), 1115-1120.

4. R. Xiang and Cs. Horvath. Fundamentals of Capillary Electrochromatography: Migration Behavior of Ionized Sample Components, *Anal. Chem.* 2002: 74, 762-770.

5. J. Leonil, V. Gagnaire, D. Molle, S. pezennec, S. Bouhallab. Application of chromatography and mass spectrometry to the characterization of food proteins and derived peptides *J. Chromatogr. A*. 2000: 881(1), 1-21.

6. H. F. Alomirah, I. Alli and Y. Konishi. Applications of mass spectrometry to food proteins and peptides. *J. Chromatogr. A*. 2000: 893, 1-21.

7. C. Bayard, F. Lottspeich. Bioanalytical characterization of proteins. *J. Chromatogr. B*. 2001: 756, 113-122.

8. M. V. Moreno-Arribas, E. Pueyo and M. C. Polo. Analytical methods for the characterization of proteins and peptides in wines. *Anal. Chem. Acta*. 2002: 458, 63-75.

9. M. Careri, F. Bianchi, C. Corradini. Recent advances in the application of mass spectrometry in food-related analysis. J. Chromatogr. A. 2002: 970, 3-64.

10. F. kvasnicka. Proteomics: general strategies and application to nutritionally relevant proteins *J. Chromatogr. B*. 2003: 787, 77-89.

11. H.Z. Wan, S. Kaneshiro, J. Frenz, J. Cacia. Rapid method for monitoring galactosylation levels during recombinant antibody production by electrospray mass spectrometry with selective-ion monitoring. *J. Chromatogr. A*. 2001: 913, 437-446.

12. International Conference on Harmonisation (ICH), Q2A: Text on Validation of Analytical Procedures, U.S. FDA Federal Register. 1995: 60, 11260.

13. International Conference on Harmonisation (ICH), Q2B: Validation of Analytical Procedures: Methodology, FDA Federal Register. 1997: 62, 27463.

14. G.A. Shabir, Step-by-step analytical methods validation and protocol in the quality system compliance industry. *J. Validation Technol.* 2004: 10, 314-324.

21 DEVELOPMENT AND VALIDATION OF A HPLC METHOD FOR THE DETERMINATION OF 9α-FLUORO-16β-METHYL-PREDNISOLONE-17-VALERATE

ABSTRACT

In this study, the development and validation of an analytical method for the assay of 9α-fluoro-16β-methyl-prednisolone-17-valerate (FMPV) using reversed-phase high-performance liquid chromatography (HPLC) is reported. The chromatographic separation is achieved with methanol-water (63:37, v/v) as mobile phase, a C_{18} column, and UV detection at 240 nm. The method was critically validated to demonstrate its selectivity, linearity, precision, accuracy, specificity, limit of detection and quantitation. The calibration curve showed good linearity (r^2 = 0.9999) over the concentration range 20 to 150 µg/mL. The mean percent relative standard deviation values for precision studies were less than 0.37%. Mean recoveries were 99.73-100.02%. The limit of detection was 3.0 µg/mL and limit of quantitation was 20.0 µg/mL. This method represents a useful protocol for routine testing of 9α-fluoro-16β-methyl-prednisolone-17-valerate drug substance.

Keywords: HPLC, 9α-fluoro-16β-methyl-prednisolone-17-valerate drug, Method development, Method validation

1. INTRODUCTION

9α-Fluoro-16β-methyl-prednisolone-17-valerate (FMPV, Figure 1) is a steroidal anti-inflammatory drug used in medicinal formulations. FMPV is an active topical corticosteroid that produces a rapid response in those inflammatory dermatoses that are normally responsive to topical corticosteroid therapy, and is often effective in the less responsive conditions such as psoriasis. It is used to treat many conditions including dermatitis, arthritis, inflammatory bowel disease, reactive airways disease, and respiratory distress syndrome in preterm infants and pruritus in corticosteroid-responsive dermatoses. FMPV is an example of a synthetic drug. It is designed to be a substitute for cortisone in relieving the symptoms of rheumatoid arthritis with fewer undesirable side effects [1]. FMPV is also chemically known as 9-fluoro-11-beta,17,21-trihydroxy-16-beta-methylpregna-1,4-diene-3,20-dione-17-valerate. In the present study, a simple reversed-phase high-performance liquid chromatographic (HPLC) assay for FMPV was developed and extensively validated. HPLC is a widespread separation technique that occupies the leading position in routine pharmaceutical analysis.

Reversed-phase chromatography is probably the most commonly used separation mechanism in liquid chromatography and consists of a non-polar stationary phase (normally octadecyl, C_{18} or octyl C_8 chains) bonded to a solid support that is generally micro particulate silica gel (non-polar). The mobile phase is polar and, therefore, the sample compounds are partitioned between the mobile and the stationary phases. The separation is normally performed using aqueous mobile phase containing different percentages of organic modifiers (e.g. methanol, ethanol, acetonitrile, or THF) to increase the selectivity between species. Solute retention is also influenced by eluent pH, which affects the dissociation level of the analyte and therefore, its partition between the mobile and stationary phases. Most of the analytical techniques for FMPV described in the literature are based on the liquid chromatographic determination of this drug in topical dosage forms, creams, lotions, and ointment formulations involve a liquid-liquid extraction, photochemical dramatization procedures [2-7], alcoholic extraction [8] or silica gel column separation [9] prior to chromatography, which are expensive and time consuming. The aim of the present work was to develop a cost effective, simple, fast, precise, and accurate HPLC method to be applied to the quantitative analysis of incoming raw materials of FMPV used in pharmaceutical formulations. The method was fully validated using step-by-step protocol [10] as a best practice [11] and guidelines [12-15].

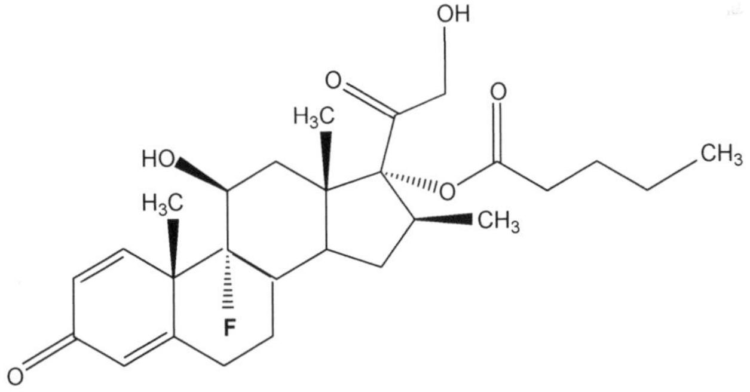

FIGURE 1 Chemical structure of 9α-fluoro-16β-methyl-prednisolone-17-valerate.

2. EXPERIMENTAL

2.1 Chemicals and Reagents

Methanol (HPLC-grade) was obtained from Merck (Darmstadt, Germany). 9α-Fluoro-16β-methyl-prednisolone-17-valerate (FMPV, 98% pure), sodium dihydrogen phosphate (NaH_2PO_4), and orthophosphoric acid (H_3PO_4) were purchased from Sigma chemicals (St. Louis, MO, USA). De-ionised distilled water was used throughout the experiment. All other reagents were of analytical grade.

2.2 HPLC System and Analytical Conditions

A PerkinElmer (Norwalk, CT) HPLC system equipped with a module LC 235C diode array detector (DAD), series 200 LC pump, series 200 autosampler and series 200 peltier LC column oven were used in this work. The data were acquired via PE TurboChrom Workstation data acquisition software using PE Nelson series 600 LINK interfaces. A mixture of methanol-water (63:37, v/v) was used as mobile phase at a flow rate of 0.8 mL/min. The mobile phase was filtered through 0.45 μm membrane filter and continuously degassed on-line. The injection volume was 10 μL and the detection wavelength was set at 240 nm. The chromatographic separation was achieved using a 100 × 3 mm, C_{18} ChromSpher polymeric octadecylsilane (ODS)-encapsulated spherical silica column with a 5 μm particle size obtained from Varian (Palo Alto, CA, USA). The separation was carried out at temperature, 25 ±0.5°C.

2.3 Preparation of the Standard and Sample Solutions

An accurately weighed amount (50 μg) of 9α-fluoro-16β-methyl-prednisolone-17-valerate, standard was placed in a 100 mL volumetric flask and dissolved in methanol (stock). Pipette 10 mL aliquot of stock solution to a second 100 mL volumetric flask, added 53 mL methanol and volume made up with 25 mM phosphate buffer (adjusted to pH 3.0 using orthophosphoric acid). Linearity experiment was performed by preparing drug substance (9α-fluoro-16β-methyl-prednisolone-17-valerate) in the range of 20-150 μg/mL.

3. RESULTS AND DISCUSSION

3.1 Chromatography

To obtain the best chromatographic conditions, different columns and mobile phases consisting of acetonitrile-water or methanol-water were tested to provide sufficient selectivity and sensitivity in a short separation time. The best signal was achieved using methanol-water (63:37, v/v) with a flow rate of 0.8 mL/min in a C_{18} analytical column. The low flow rate and the short run time resulted, comparatively, in lower consumption of the mobile phase solvents with a better cost effective relation. 10 μL of solutions were injected automatically into the column. The optimal wavelength for FMPV detection was established using two UV absorbance scans over the range of 190 to 400 nm, one scan of the mobile phase, and the second of the analyte in the mobile phase. It was shown that 240 nm is the optimal wavelength to maximize the signal. A typical chromatogram obtained by the proposed HPLC method, is shown in Figure 2. The low retention time of 4.41 min allow a rapid determination of the drug, which is an important advantage for the routine analysis.

To evaluate the quantitative nature of the analytical method, a series of samples with different amounts of FMPV were run to investigate the best assay concentration. Using a C_{18} column, best concentration was assessed by injecting six standards of drug in the range of 10 to 200 µg/mL. The integrated peak areas (µVs) were plotted versus amount injected. The calibration curve was found to be linear from concentration range 20 to 150 µg/mL with correlation coefficient of 0.9999. On the bases of this data, the best concentration (50 µg/mL) was chosen as a working concentration for the assay.

System suitability testing was performed to determine the accuracy and precision of the system from six replicate injections of a solution containing 50 µg FMPV /mL. The percent relative standard deviation (%RSD) of the retention time (min) and peak area were found to be less than 0.29%. The retention factor (also called capacity factor, k) was calculated using the equation $k = (t_r / t_0) -1$, where t_r is the retention time of the analyte and t_0 is the retention time of an unretained compound; in this study, t_0 was calculated from the first disturbance of the baseline after injection and capacity factor value was obtained 9.32 for FMPV peak. The separation factor (α) was calculated using the equation, $\alpha = k_2 / k_1$ where k_1 and k_2 are the retention factors for the first and last eluted peaks respectively. The separation factor for FMPV peak was 2.18 obtained. The plate number (also known as column efficiency, N) was calculated as $N = 5.54 \ (t_r / w_{0.5})^2$ where $w_{0.5}$ is the peak width at half peak height. In this study, the theoretical plate number was 2812. Resolution is calculated from the equation $R_s = 2(t_2 - t_1) / (t_{w1} + t_{w2})$. Where t_1 and t_2 are retention times of the first and second eluted peaks, respectively, and t_{w1} and t_{w2} are the peak widths. The resolution for FMPV peak was > 2.0. The asymmetry factor (A_s) was calculated using the US Pharmacopoeia (USP) method. The peak asymmetry value for each FMPV peak was 1.07.

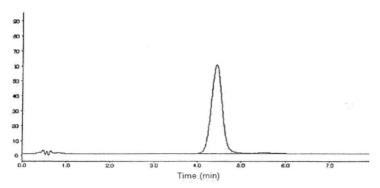

FIGURE 2 Chromatogram obtained for a 9α-fluoro-16β-methy-lprednisolone-17-valerate standard solution.

Robustness studies were also performed in the method development phase applying the experimental design as shown in Table 1. A sample of FMPV was prepared at working concentration

(50 µg/mL) and assayed using the experimental design with eight-test combination for seven different chromatographic parameters as shown in Table 2. For each parameter, four combinations of (AAAA) and four combinations of (aaaa) were studied. The actual value of each parameter (V_A - V_G) (Table 2) shows which parameter has a dominant influence on the developed analytical method. In all cases good separations of FMPV was always achieved, indicating that the analytical method remained selective for FMPV drug substance under the optimized conditions.

TABLE 1 Experimental design for robustness study

Test parameter	1	2	3	4	5	6	7	8
A/a	A	A	A	A	a	a	a	a
B/b	B	B	b	b	B	B	b	b
C/c	C	c	c	c	C	c	C	c
D/d	D	D	d	d	d	d	D	D
E/e	E	e	E	e	e	E	e	E
F/f	F	f	f	F	F	f	f	F
G/g	G	g	g	G	g	G	G	g
Results	s	t	u	v	w	x	y	z

TABLE 2 Chromatographic parameter for robustness study

Parameter	Test conditions 1	Test conditions 2	Differences
Analytical column	A = column C-18	a = column C-18	$V_A = \frac{1}{4}(s+t+u+v) - \frac{1}{4}(w+x+y+z) = A - a$
Sample solvent	B = Buffer/methanol	b = Mobile phase	$V_B = \frac{1}{4}(s+t+w+x) - \frac{1}{4}(u+v+y+z) = B - b$
Temperature	C = 20 °C	c = 30 °C	$V_C = \frac{1}{4}(s+u+w+y) - \frac{1}{4}(t+v+x+z) = C - c$
Flow rate	D = 0.6 ml/min	d = 1.0 ml/min	$V_D = \frac{1}{4}(s+t+y+z) - \frac{1}{4}(u+v+w+x) = D - d$
Wavelength	E = 238 nm	e = 242 nm	$V_E = \frac{1}{4}(s+u+x+z) - \frac{1}{4}(t+v+w+y) = E - e$
Mobile phase	F = 61% methanol	f = 65 % methanol	$V_F = \frac{1}{4}(s+v+w+z) - \frac{1}{4}(t+u+x+y) = F - f$
Solubility stability	G = 24 h	g = 48 h	$V_G = \frac{1}{4}(s+v+x+y) - \frac{1}{4}(t+u+w+z) = G - g$

3.2 Validation of the Method

3.2.1 Linearity

Linearity was studied using six solutions in the concentration range 20-150 µg/mL and each one injected in duplicate. The regression equation was found by plotting the peak area (y) versus the FMPV concentration (x) expressed in µg/mL. The correlation coefficient (0.9999) obtained for the regression line demonstrates that there is a strong linear relationship between peak area and concentration of FMPV (Table 3).

3.2.2 Accuracy

The accuracy of an analytical method is determined by how close the test results obtained by that method come to the true value. It can be determined by application of the analytical procedure to an analyte of known purity (for the drug substance) or by recovery studies, where a known amount of standard is spiked in the placebo (for drug product). In the present study, a number of different solutions were prepared with a known added amount of drug substance and injected in triplicate. Percent recoveries of response factor (area and concentration) were calculated as can be seen in Table 3, and it is evident that the method is accurate within the desired range.

3.2.3 Precision studies

The precision of the analytical method, reported as %RSD, was estimated by measuring repeatability (intra-day precision) on ten replicate injections at 100% test concentration.

Intermediate precision (inter-day variation) was demonstrated by two analysts using two HPLC systems, and evaluating the relative peak area percent data across the two HPLC systems at three concentration levels (40, 60, and 80%). The %RSD values presented in Table 3 were less than 0.37% in all cases, and illustrated the good precision of the chromatographic method. Figure 3 demonstrates day 1 and day 2 variations.

TABLE 3 Method validation results

Validation step	Concentration as % of 50 µg/mL	Results
Linearity ($k = 6$, $n = 2$)	20-150	$y = 26679x - 115361$ ($r^2 = 0.9999$)
Accuracy		
(%Recovery, %RSD, $n = 3$)	40	100.02 ($\pm$0.03)
	80	99.88 ($\pm$0.10)
	150	99.73 ($\pm$0.10)
Repeatability		
(Peak area, %RSD, $n = 10$)	50	0.27
Intermediate precision ($n = 3$)		
(Day 1, %RSD)	40	0.31
	60	0.34
	80	0.29
(Day 2, %RSD)	40	0.28
	60	0.36
	80	0.33
LOD		(s/n = 3.2), 3 µg/mL
LOQ ($n = 6$)		(s/n = 10.2), 20 µg/mL
Stability		
(%Change in response factors)	50	0.15

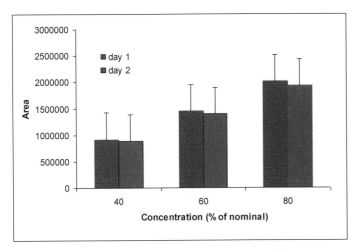

FIGURE 3 Typical graph obtained after analysis of FMPV to demonstrate intermediate precision variation studied over 2 consecutive days.

3.2.4 Specificity

Forced degradation studies were performed to evaluate the specificity of FMPV under four stress conditions (heat, UV light, acid, base). Solutions of FMPV were exposed to 50°C for 1 h, UV light using a Mineralight UVGL-58 light for 24h, acid (1M hydrochloric acid) for 24 h, and base (1M sodium hydroxide) for 4 h. A summary data of the stress results is shown in Table 4, which showed no changes in retention times of each FMPV by peak purity analysis on a DAD UV detector and, therefore, confirms the specificity of the method.

TABLE 4 Validation results obtained for the assay of FMPV under stress conditions

Stress conditions	Sample treatment	Retention time (min)	%Recovery	Peak area (μVs)
Reference	Fresh solution	4.41	99.99	1068056
Acid	1M HCl for 24 hour	4.40	99.97	1056255
Base	1M NaOH for 4 hour	4.41	99.98	1042723
Heat	50°C for 1 hour	4.41	99.96	1038726
Light	UV Light for 24 hour	4.40	99.97	1066276

3.2.5 Limits of detection and quantitation

The limit of detection (LOD) and limit of quantitation (LOQ) tests for the procedure were performed on samples containing very low concentration of analyte. LOD is defined as the lowest concentration of analyte in a sample that can be detected above baseline noise. It is expressed as a concentration at a specified signal-to-noise (s/n) ratio, typically, three times the noise level. LOQ is defined as the lowest concentration of analyte in a sample that can be reproducibly quantitated above the baseline noise that gives s/n > 10. The LOD for FMPV was 3.0 μg/mL and s/n was >3.2. The LOQ was (s/n 10.2) 20.0 μg/mL and %RSD for six injections was 0.31% (Table 3).

3.2.6 Stability of analytical solutions

Three standard solutions (50 μg/mL) were chromatographed immediately after preparation and then reassayed after storage at room temperature (22°C ±1°C) for 48 h. The results given in Table 3 show that there was no significant change (< 1% response factor) in FMPV concentration over this period.

4. CONCLUSION

A new, fast, simple, cost effective and accurate reversed-phase HPLC method for the determination of FMPV drug substance has been developed and extensively validated. The results showed that the method is selective; no significant interference peak was detected; accurate, with the FMPV recoveries of 99.73-100.02%, robust and reproducible with the %RSD less than 0.37% in all cases. The method was sensitive; as little as 3.0 µg/mL could be detected with the LOQ of 20.0 µg/mL. The suggested technique can be used in quality control for release of incoming raw material of 9α-fluoro-16β-methyl-prednisolone-17-valerate drug substance used in pharmaceutical products.

REFERENCES

1. Seyhan, N.E. Organic chemistry, D.C. Heath and Company: USA, 1989; 477 pp.
2. Bailey, F. and Brittain, P.N. The quantitative determination of fluocinolone acetonide and acetonide acetate in formulated products by HPLC. *J. Pharm. Pharmacol.* 1972, *24*, 425-428.
3. Smith, E.W., Haigh, J.M. and Kanfer, I. A stability-indicating HPLC assay with on-line clean up for betamethasone 17-valerate in topical dosage forms. *Int. J. Pharm.* 1985, *27*, 185-192.
4. Burgess, C. Rapid RP-HPLC analysis of steroid products. *J. Chromatogr.* 1978, *149*, 233-240.
5. Li Wan P., Irwin, W.J. and Yip, Y.W. HPLC assay of betamethasone 17-valerateand its degradation products. *J. Chromatogr.* 1979, *176*, 399-405.

6. Metha, A.C., Calvert, R.T. and Ryatt, K.S. Betamethasone 17-valerate. An investigation into its stability in betnovate after diluting with emulsifying ointment; quantitation of degradation products. *Br. J. Pharm. Prac.* 1982, *3*, 10-13.

7. Pietra, A.M.D., Andrisano, V., Gotti, R and Cavrini, V. On-line post-column photochemical derivatisation in LC-DAD analysis of binary drug mixtures. *J. Pharm. Biom. Anal.* 1996, *14*, 1191-1199.

8. United States Pharmacopoeia 29-National Formulary 24, United States Pharmacopeial Convention, Rockville, MD, 2006, p. 996

9. Tokunaga, H., Kimura, T. and Kawamura, J. Determination of glucocorticoids by LC. Application to ointments and a cream containing cortisone acetate, dexamethasone acetate, fluorometholone and betamethasone valerate. *Chem. Pharm. Bull.* 1984, *32*, 4012-4016.

10. Shabir, G.A. Step-by-step analytical methods validation and protocol in the quality system compliance industry. *J. Validation Technol.* 2004, *10*, 314-324.

11. Shabir, G.A., Lough, W.J., Shafique, A. A., Bradshaw, T.K. Evaluation and application of best practice in analytical method validation. *J. Liq. Chromatogr. Rel. Technol.* 2007, *30*, 311-333.

12. Shabir, G.A. Validation of HPLC methods for pharmaceutical analysis: Understanding the differences and similarities between validation requirements of the U.S. Food and Drug Administration, the U.S. Pharmacopoeia and the International Conference on Harmonization. *J. Chromatogr. A* 2003, *987*, 57-66.

13. International Conference on Harmonization (ICH), Q2A: Text on validation of analytical procedures, March 1995.

14. International Conference on Harmonization (ICH), Q2B: Validation of analytical procedures: Methodology, May 1997.

15. United States Pharmacopoeia 29-National Formulary 24, General chapter 1225, Validation of compendial procedures. United States Pharmacopeial Convention, Rockville, MD, 2006, p. 3050.

22 DETERMINATION OF 11β,17α, 21-TRIHYDROXYPREGN-4-ENE-3,20-DIONE RESIDUES ON MANUFACTURING EQUIPMENT SURFACE USING HPLC

ABSTRACT

A new high-performance liquid chromatography (HPLC) method for the assay of 11β,17α, 21-trihydroxypregn-4-ene-3,20-dione residues in swab samples collected from pharmaceutical manufacturing equipment surfaces has been developed and validated. Chromatographic separation on a C_{18} column was achieved using methanol-water (55:45, v/v) as the mobile phase at a flow rate of 2 mL/min. The peak was detected using a UV detector set at 254 nm. The calibration curve was linear over a concentration range from 0.40-1.60 µg/mL with correlation coefficient of 0.9999. The mean percent relative standard deviation values for intra- and inter-day precision studies were less than 0.48%. Mean recoveries were between 98.70-99.81%. The limits of detection and quantitation were determined to be 0.15 and 0.40 µg/mL respectively. The method was successfully applied to the assay of actual swab samples collected from the equipment surfaces.

Keywords: HPLC, Method validation, 11β,17α, 21-trihydroxypregn-4-ene-3,20-dione, Residues in swab samples, Pharmaceutical manufacturing equipment surfaces

1. INTRODUCTION

11β,17α, 21-Trihydroxypregn-4-ene-3,20-dione (THPD, Figure 1) is a steroid hormone produced by the adrenal cortex and synthesized for medical use. It has anti-inflammatory, glucocorticoid, and sodium-retaining (mineralocorticoid) properties. It is used clinically to reduce the pain and inflammation of various conditions, including rashes, hemorrhoids, arthritis, and inflammatory bowel disease. It also is used as steroid replacement therapy in patients with adrenocortical deficiency. 11β,17α, 21-Trihydroxypregn-4-ene-3,20-dione is also known as hydrocortisone and its molecular weight is 362.47. It has been manufactured over the past fifty years using fungi such as *rhizopus arthpous* or *aspergillus niger*. These achieve the necessary step of β-hydroxylation, while *cochliobolas iunatus* was relied upon for the α-hydroxylation.

Good manufacturing practice dictates that the equipment necessary to manufacture pharmaceuticals must be maintained in a clean and orderly manner [1] In many cases, especially in a research and development plant, the same equipment may be used for processing different products and, in order to avoid cross contamination of the following pharmaceutical product, an adequate cleaning procedure is essential. The cleaning procedure validation describes responsibilities, facilities, cleaning strategies, analytical strategies and residue limit justifications. The cleaning validation consists therefore in two separate activities: the first is the development and validation of the cleaning procedure that is used to remove drug residues from the manufacturing surfaces and the second consists in developing and validation the methods used to quantify residuals from surfaces that are used in the manufacturing environment. To control the effectiveness of cleaning, the analytical method should be selective for the substance considered and has to provide a sufficient sensitivity since the concentration levels of residues are generally low. The objective of the analytical method consists in controlling that the contaminants can be removed from the equipment surface. It is therefore necessary to ensure that contaminants can be recovered from the equipment surface and to determine the level as well as the consistency of recovery.

The sampling is therefore a very important parameter since the conclusions of the cleaning procedure are based on sample results. Indeed, a negative test may be due to a deficient sampling technique. According to the Food and Drug Administration (FDA) guide, [1] two different method of sampling are generally admitted for performing a cleaning control: the direct surface sampling, using the swabbing technique and the indirect sampling based on the analysis of solutions used for rinsing the equipment (rinse method).

The rinse method occurs after the cleaning has been completed and allows the sampling of large surfaces and of inaccessible systems. Moreover, systems that cannot be routinely disassembled can be sampled and evaluated using this technique. However, it must be taken into account that the residue or contaminant may be insoluble or may be physically occluded in the equipment [1].

On the other hand, the swabbing method is a clearly more direct way of sampling. Using this method, areas hardest to clean and which are reasonably accessible can be evaluated, leading to establishing a level of contamination per given surface are. Moreover, residues that are 'dried out' or insoluble can be sampled by physical removal. As the swab sampling does not cover the entire equipment. It is important to define with care the sampling sites. Moreover, due to the nature of this method, it is of great importance to evaluate carefully the material to be used (swab, solvents etc) and to determine the efficiency of the sampling (recovery). The main objective of this paper is to propose a selective and validated high-performance liquid chromatography (HPLC) method for the assay of residual levels of THPD using swab sampling method.

Recently there have been a number of reports dealing with various analytical methods for the determination of THPD, such as UV spectrophotometry [2, 3], reverse-phase chromatography [4-7], micellar electrokinetic chromatography [8, 9] and mass spectrometry [10] as well. However, HPLC assay method has so far not been reported for determination of residual THPD in cleaning validation swab samples. The aim of this research was to develop and critically validate a cleaning method for the determination of residual THPD in swab samples collected by swabbing manufacturing equipment vessel's surfaces used in pharmaceutical manufacture of drug products containing THPD as an active component. The present report describes a method development and validation for the analysis of residual THPD extracts using HPLC. The validation of the method was performed using validated equipment in accordance with established and accepted criteria [11, 12] and step-by-step approaches and best practices [13-15].

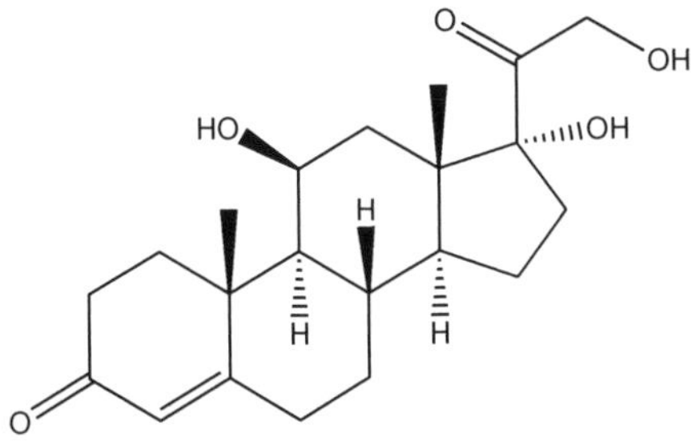

FIGURE 1 Chemical structure of 11β,17α, 21-trihydroxypregn-4-ene-3,20-dione.

2. EXPERIMENTAL

2.1 Chemicals and Reagents

All chemicals and reagents were of the highest purity. HPLC-grade methanol was obtained from Merck (Darmstadt, Germany). 11β,17α, 21-Trihydroxypregn-4-ene-3,20-dione (THPD, 98% pure) was purchased from Sigma chemicals (St. Louis, MO, USA). De-ionised distilled water was used throughout the experiment. Swabs were obtained from Baxter Products Division, McGaw Park, IL.

2.2 Equipment and Chromatographic Conditions

High-performance liquid chromatographic system (Perkin Elmer, Norwalk, CT, USA) equipped with a module LC 235C diode array detector (DAD), series 200 LC pump, series 200 autosampler and series 200 peltier LC column ovens were used in this work. The data were acquired via PE TurboChrom Workstation data acquisition software using PE Nelson series 600 LINK interfaces. The second HPLC system was used in the validation study during intermediate precision was also a Perkin Elmer.

Analytical conditions were optimized through the LC system using C_{18} µ-Bondapak column (300 mm × 3.9 mm, 10-µm particle size) WATERS® (Milford, MA). The mobile phase was consisted of a mixture of methanol-water (55:45, v/v). The flow rate of the eluent was 2 mL/min and the column temperature was maintained at 25°C. The injection volume was 20 µL and the detection wavelength was set at 254 nm.

2.3 Standard Preparation

An accurately weighed amount (1 mg) of THPD reference standard was placed in a 100 mL volumetric flask and dissolved in methanol (stock). A 5 mL aliquot of stock solution was diluted to 50 mL in mobile phase, yielding a final concentration of 1 µg/mL.

2.4 Preparation of Calibration Solutions

A stock standard solution of THPD was prepared in methanol at 10 μg/mL. This solution was then diluted with mobile phase in order to obtain the calibration solutions at concentrations of 0.40, 0.60, 0.80, 1.0, 1.20 and 1.60 μg/mL.

2.5 Sample Solution for Determination of Inter and Intra-day Precision

The sample solution was prepared for repeatability (inter-day precision) studies as described in section 2.3 and ten replicate injections were made into the chromatograph. The sample solutions for intra-day precision were prepared as follows: a stock standard solution of THPD was prepared in methanol at 10 μg/mL. This solution was then diluted with mobile phase in order to obtain the intra-day precision solutions at concentrations of 0.40, 1.0 and 1.60 μg/mL. These solutions were injected in triplicate into the chromatographic system.

2.6 Sample Preparation for Recovery Test

Swabs were soaked with methanol and wiped 20 cm^2 stainless steel surface area. These swabs were placed into a 100 mL HDPE bottles. A 50 mL methanol was added to each bottle. The known amounts of THPD standard (0.402, 1.000, and 1.601 μg/mL) were added to each swab sample bottles, respectively. Each bottle was capped and swabs were extracted using mechanical shaker for 15 min. The extract was transferred to a 100 mL volumetric flask and then diluted in mobile phase. The extract was filtered through a 0.45 μm membrane filter and injected into the chromatograph.

2.7 Sample Preparation

Swabs were soaked with methanol (swabbing solvent) and wiped 20 cm² stainless steel surface area first in a horizontal manner insuring that the total surface is wiped, and then in a vertical way, starting from the outside towards the centre. Overlapping the same surface is acceptable. (Note: vinyl, powder free, gloves must be worn to avoid interferences). These swabs were placed into a 100 mL HDPE bottles. A 50 mL methanol was added to each bottle. Each bottle was capped and swabs were extracted using mechanical shaker for 15 min. The extract was transferred to a 100 mL volumetric flask and then diluted in mobile phase. The extract was filtered through a 0.45 μm membrane filter and injected into the chromatograph.

3. RESULTS AND DISCUSSION

3.1 Method Development

The liquid chromatographic analysis of THPD was carried out in the isocratic mode using a final optimised composition of methanol-water (55:45, v/v) as mobile phase. Preliminary experiments were performed to select the analytical column most suitable for the separation of THPD. Chromatography of hydrophobic compounds on reversed-phase packing material has been associated with asymmetric peak shapes due to adsorption reactions of the solute with non-bonded silanol sites on the stationary phase. Stationary phase which possess a maximum surface coverage by C_8 and C_{18} groups and which have been exhaustively end-capped have been shown to results in improved chromatography of such hydrophobic compounds. Two C_{18} columns, μ-Bondapak and Partisil-10-ODS-3 were tested. The μ-Bondapak analytical column gave the best separation probably due to its higher coverage of silanol sites. The column was equilibrated with the mobile phase flowing at 2 mL/min for 30 min prior to injection. 20 μL of standard and sample solutions were injected automatically into the column. The optimal wavelength for THPD detection was established using two UV absorbance scans over the range of 190 to 400 nm, one scan of the mobile phase, and the second of the analyte in the mobile phase. It was shown that 254 nm is the optimal wavelength to maximise the signal (Figure 2). Chromatograms of the THPD gave good peak shape (Figure 3). The retention time for THPD was 5.2 min. To evaluate the quantitative nature of the analytical method, a series of samples with different amounts of THPD were run to investigate the best assay concentration. Using a C_{18} μ-Bondapak column, best concentration (1 μg/mL) was assessed by injecting six standards of THPD in the range of 0.01-2.5 μg/mL. The integrated peak areas were plotted versus amount injected. The calibration curve was found to be linear from concentration range 0.40-1.60 μg/mL with a correlation coefficient of 0.9999. On the bases of these data, the middle concentration (1 µg/mL) of the linearity studied was chosen for the assay.

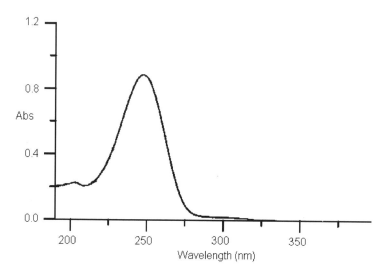

FIGURE 2 DAD UV spectra of the middle of the peak corresponding to the RT of main component of THPD.

Robustness studies were also performed in the method development optimization phase that showed 20 µL injection volume was reproducible and the peak response was significant at the analytical concentration (1 µg/mL) chosen. During robustness study, a number of chromatographic parameters were investigated such as flow rate, column temperature, mobile phase composition, and column from different suppliers. The robustness measurement data is presented in Table 1. In all cases good separations of THPD was always achieved, indicating that the analytical method remained selective for THPD component under the tested conditions.

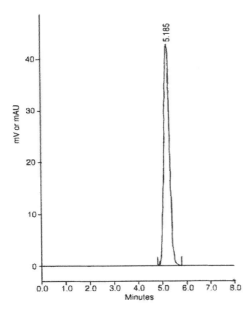

FIGURE 3 Chromatogram for the separation of THPD standard.
Chromatographic conditions: Column C_{18} µ-Bondapak (300 mm x 3.9 mm, 10-µm particle size); Mobile phase, isocratic elution with methanol/water (55:45, v/v); Flow rate 2 mL/min; Column temperature 25°C; Sample concentration 1 µg/mL; Injection volume 20 µL.

The stability was studied on standard solutions at concentration of 0.40, 1.0, and 1.60 µg/mL. These solutions were stored for 7 days at 25°C, away from direct sunlight and reanalyzed by following the proposed method. The results given in Table 2 show that there was no significant change (<1% response factor) in THPD concentration over this period.

Table 1 Chromatographic parameter for robustness study of THPD

Factor	Experimental domain	Optimized conditions	RT (min)	Area (μVs)
HPLC columns (different suppliers)	C18 μ-Bondapak and Partisil-10-ODS-3	C18 μ-Bondapak	5.19	6658622
Percent organic solvent	52-58	55	5.19	6654581
Flow rate (mL/min)	1.8-2.2	2.0	5.19	6662486
Injection volume (μL)	18-22	20	5.20	6645865
Column temperature (°C)	23-27	25	5.19	6598845

3.2 Method Validation

3.2.1 Linearity and range

Linearity was studied in the concentration range 0.40-1.60 μg/mL (K = 6; n = 3) and the following regression equation was found by plotting the peak area (y) versus the THPD concentration (x) expressed in μg/mL:

$$y = 149157x + 32695 \ (r^2 = 0.9999) \qquad (1)$$

The determination coefficient (r^2) obtained (Table 2) for the regression line demonstrates the excellent relationship between peak area and the concentration of THPD.

TABLE 2 Validation results obtained for the HPLC assay of THPD

Validation criterion	Concentration range (µg/mL)	Results
Linearity (k = 6; n = 3)	0.40-1.60	y = 149157x + 32695 (r^2 = 0.9999)
Repeatability (RT, area, height, %RSD, n = 10)	1.0	0.10 0.15 0.05
LOD		0.15 µg/mL
LOQ		0.40 µg/mL
Stability (% Change in response factor)	0.40 1.0 1.60	0.24 0.18 0.36

3.2.2 Precision (repeatability and intermediate precision)

The precision of the chromatographic method, reported as percent RSD, was estimated by measuring repeatability on ten replicate injections at 100% test concentration (1 µg/mL). Repeatability is also called intra-assay or within-run precision. Repeatability was calculated using equation (2). The RSD values for retention time (min), peak areas and peak height were 0.10, 0.15 and 0.05, respectively (Table 2).

$$\%RSD = \frac{sample\ replicate\ sd}{sample\ replicate\ mean} \ X\ 100 \qquad (2)$$

The intermediate precision (inter-day variation) was demonstrated by two analysts using two HPLC systems on different days and evaluating the peak area data across the HPLC systems at three concentration levels (0.40, 1.0, and 1.60 µg/mL) that cover the assay method range 0.40 to 1.60 µg/mL for THPD. The RSD across the HPLC systems and analysts were calculated using equation (3) from the individual peak area total mean values at the 0.40, 1.0, and 1.60 µg/mL of the test concentration.

$$\%RSD = \frac{total\ mean\ sd}{total\ mean} \times 100 \qquad (3)$$

The RSD values for both instruments and analysts were < 0.48% (Table 3) for THPD and illustrated an excellent precision of the analytical method.

TABLE 3 Intermediate precision studies data for THPD

Injection no.	Analyst 1, day 1, HPLC system 1			Analyst 2, day 2, HPLC system 2		
	0.40[a]	1.00	1.60	0.40	1.00	1.60
1	6652764[b]	6657943	6659534	6657654	6659845	6659598
2	6662523	6657244	6653696	6673266	6654498	6658711
3	6662399	6662976	6667622	6599436	6665689	6659878
Mean	6659229	6659388	6660284	6643452	6660011	6659396
RSD (%)	0.47	0.34	0.25	0.42	0.37	0.22

[a]Concentration (µg/mL); [b]Peak area (µV)

3.2.3 Accuracy

The accuracy of the procedure was assessed by comparing the analyte amount determined versus the known amount spiked in swab samples at three different concentration levels (0.402, 1.000, and 1.601 µg/mL) with three replicates for each concentration level studied. The accuracy defined as mean percent associated with an interval of confidence (IC; $p = 0.05$) shows that the HPLC method developed for the determination of THPD can be considered as accurate within the concentration range studied. The recovery data is presented in Table 4.

TABLE 4 Results obtained in the recovery of THPD standard added to sample solutions and analyzed by the proposed HPLC method

Amount added (µg/mL)	Amount recovered (µg/mL)	Recovery (%, $n = 3$)	RSD (%, $n = 3$)	Mean recovery (%)
0.402	0.398	99.00	0.24	
1.000	0.987	98.70	0.35	99.17
1.601	1.598	99.81	0.18	

3.2.4 Specificity

A method is specific for a particular compound if the measured response is due to that compound and due to no other compound present in the sample. Injections of the extracted placebo were performed to demonstrate the absence of interference from any other materials with the elution of the THPD at same retention time. These results demonstrate that there was no interference from the other compounds in the formulation and, therefore, confirm the specificity of this method (Figure 4).

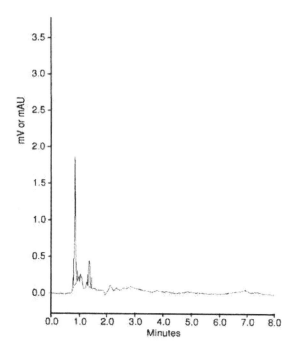

FIGURE 4 Separation obtained from placebo formulation without adding THPD. Chromatographic conditions are similar to that of Figure 3.

3.2.5 Limits of detection and quantitation

The limit of detection (LOD) and limit of quantitation (LOQ) of THPD were estimated from the intercept ($\bar{a}$) of the regression line and the corresponding residual standard deviation ($S_{y/x}$). [16] The responses at the LOD and LOQ were estimated by the following expressions respectively.

$$f(\text{LOD}) = \bar{a} + 3S_{y/x} \qquad (4)$$
$$f(\text{LOQ}) = \bar{a} + 10S_{y/x} \qquad (5)$$

Applying this method, LOD and LOQ for THPD were found to be 0.15 and 0.40 μg/mL, respectively (Table 2).

4. APPLICATION OF THE METHOD

The applicability of the developed and critically validated method was evaluated: swab samples from different locations within the manufacturing equipment were submitted to the quality control laboratory for the analysis of residual THPD. These samples were prepared and analysed applying the proposed method. A typical chromatogram obtained for a THPD cleaning validation swab sample obtained from a location within the equipment is presented in Figure 5. It is evident from chromatogram in Fig. 5, traces of THPD residual was not found at the same retention time of THPD (RT = 5.19) and confirmed the equipment cleaning validation process.

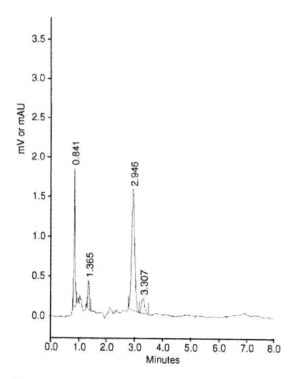

FIGURE 5 Separation obtained for a representative swab sample collected from the manufacturing equipment surfaces show absence of elution of THPD (RT = 5.19 min). Chromatographic conditions are similar to that of Figure 3.

5. CONCLUSION

A new HPLC cleaning validation method for the determination of residual levels of THPD in swab samples collected by wiping different surfaces from the manufacturing equipment vessels has been developed and validated. The proposed method was demonstrated to be simple, sensitive, linear, accurate, precise and robust in the concentration range studied. The sampling procedure selected consists in using swabs previously moistened with methanol. It should be noted that the swab sample must be performed immediately after the completion of the cleaning process to avoid under estimation of the residues of drug remaining on the equipment surface. The proposed method was successfully implemented in the quality control laboratory to the routine analysis of actual swab samples collected from equipment surfaces.

REFERENCES

1. Guide to Inspections Validation of Cleaning Processes. (1993). Reference Material for FDA Investigators and Personnel, Food and Drug Administration, Washington, DC, pp. 1-6.
2. Blanco, M.; Coello J.; Iturriaga, H.; Naspoch, S.; Villegas, N. Spectrophotometric determination of hydrocortisone acetate in a pharmaceutical preparation by use of partial least- squares regression. Analyst 1999, *124*, 911- 915.
3. Amin, A. S. Colorimetric microdetermination of some corticosteroid drugs using indophenol as chromophoric reagent. *Anal lett.* 1996, *29(9)*, 1527-1537.
4. Hajkova, R.; Solich, P.; Dvoak, J.; Sicha, J. Simultaneous determination of methylparaben, propylparaben, hydrocortisone acetate and its degradation products in a topical cream by RP-HPLC. *J. Pharm. Biomed. Anal.* 2003, *32(4-5)*, 921-927.
5. Galmier, M. J.; Beissac, et.al. Validation of a RP-HPLCmethod for the determination of hydrocortisone phosphate disodium in a gel formulation. *J. Pharm. Biomed. Anal.* 1999, *20*, 405-409.
6. Doppenschmitt, S.A.; Scheidel, B.; Harrison.Simultaneous determination of triamcinolone acetonide and hydrocortisone in human plasma by hplc. *J. Chromatogr. B: Biomed. Sci.* Appl. 1996, *682(1)*, 79-88.

7. Grippa, E.; Santini, et.al. Simultaneous determination of hydrocortisone, dexamethasone, indomethacin, phenylbutazone and oxyphenbutazone in equine serum by HPLC. *J. Chromatogr. B: Biomed. Sci. Appl.* 2000, *738 (1)*, 17-25.

8. Rao, L.V.; Petersen. J.R.; Bissell, et.al. Development of a urinary free cortisol assay using solid-phase extraction-capillary electrophoresis. *J. Chromatogr. B* 1999, *730*, 123-128.

9. Lemus, G.J.M.; Pérez, A.J. Determination of hydrocortisone, polymyxin B and Zn-bacitracin in pharmaceutical preparations by micellar electrokinetic chromatography. *Anal. Bioanal. Chem.* 2003, *375*, 617-622.

10. Bevalot, F.; Gaillard. Analysis of corticosteroids in hair by liquid chromatography–electrospray ionization mass spectrometry. *J.Chromatogr. B, Biomed. Appl.* 2000, *740(2)*, 227-236.

11. International Conference on Harmonization. (2005). Validation of Analytical Procedures: Text and Methodology, Q2(R1).

12. U.S. Pharmacopeia. (2007). General Chapter 1225, Validation of Compendial Procedures, USP Convention Inc. Rockville, MD, pp. 680.

13. Shabir, G.A. Validation of HPLC Methods for Pharmaceutical Analysis: Understanding the differences and similarities between validation requirements of FDA, USP and ICH. *J. Chromatogr. A* 2003, *987*, 57-66.

14. Shabir, G.A.; Lough, W.J.; Arain, S.A.; Bradshaw, T.K. Evaluation and application of best practice in analytical method validation. *J. Liq. Chromatogr. Rel. Technol.* 2007, *30*, 311-333.

15. Shabir, G.A. Step-by-step analytical methods validation and protocol in the quality system compliance industry. *J. Validation Technol.* 2004, *10*, 314-324.

16. Miller, J.C.; Miller, J.N. (1993). Statistics for Analytical Chemistry, third ed., Ellis Horwood, New York.

23 DEVELOPMENT AND VALIDATION OF A HPLC METHOD FOR THE DETERMINATION OF NICOTINAMIDE ADENINE DINUCLEOTIDE: APPLICATION TO STABILITY STUDIES IN BUFFERED SOLUTIONS AND DRY TEST STRIPS

ABSTRACT

A reversed-phase high performance liquid chromatographic (RP-HPLC) method for nicotinamide adenine dinucleotide (NAD) has been developed and validated. The method was then applied to stability studies of NAD in aqueous buffered solutions and commercial dry test strips (Optium Plus™, Abbott Laboratories) for assaying blood glucose. NAD was resolved from its decomposition products on a XTerra C_{18} column (150 mm × 4.6 mm, 5 μm) using a mobile phase composed of a mixture of methanol and 20 mM potassium phosphate buffer pH 8.0 (5:95 v/v) at a flow rate of 1 mL/min with UV detection at 260 nm. The chief decomposition products were identified as ADP-ribose and nicotinamide, the amount of which depended upon the identity of the buffer. The solid-state stability of NAD formulated in Optium Plus™ test strips containing BES buffer at pH 7.0 was determined to be acceptable; > 80% residual NAD after 3 months at 50°C.

Keywords: NAD, HPLC, Method development, Method validation, Stability studies

1. INTRODUCTION

Nicotinamide adenine dinucleotide (NAD) is an essential cofactor for a large number of dehydrogenase enzymes, many of which have diagnostically important substrates such as glucose, lactate, hydroxybutyrate, ethanol, cholesterol etc. As such, the stability of NAD in solution and solid preparations for assay kits and test strips is of great interest to the diagnostic industry. There are many examples in the patent literature [1-8] of stabilised NAD solution formulations attesting to the high level of attention given to this area. Cofactors have been stabilised in solution through the use of NAD derivatives [1] and various additives / buffers such as zwitterionic buffers [2], thiols [3], sulfite [4], hydroxylamines [5], heavy metal ions [6], polyalcohols [7], proteins [8]. In addition, the stability of NAD has been enhanced by coupling to dextran [9, 10] and polyethylene glycol [9]. One stability study for NAD in solution pharmaceutical preparations has been conducted [11]. However, the majority of these studies have relied upon UV absorbance at 340 nm as an indirect measure of NAD stability since most are connected with spectroscopic kinetic assays involving the determination of NADH formation rates. The specific method of UV detection of NAD following HPLC has seen no use for stability studies, as far as we are aware, but has been applied in other areas such as the analysis of body fluids and tissues [12-15]. The question of cofactor stability also has some significance for the use of NAD-dependent enzymes in organic synthesis; this has been discussed in a review [16].

Early work [17] established that the first step in the decomposition of NAD is the loss of nicotinamide followed by slower cleavage of ADP-ribose to form adenosine monophosphate (AMP) and in some cases [18] adenosine diphosphate (ADP). The hydrolysis of NAD is dependent on pH, the molecule being relatively stable in acid media but not alkaline solutions [16]. In contrast, its reduced counterpart NADH displays the reverse pH-stability profile [16]. Control of pH via the use of buffers is therefore important in formulating stable assay reagent solutions containing NAD. However, NAD is known to be reactive towards nucleophiles such as sulfite [19], dithionite [20], cyanide [21], thiols [22], carbonyl compounds [23] and hydroxylamine [24] as well as buffers [25, 26]. Furthermore, it has been reported that adducts of NAD with nucleophiles can act as enzyme inhibitors. The decomposition profile of NAD under various conditions was the subject of much discussion [27] during the 1960s and 1970s due to the formation of several species [27, 28] which inhibited lactate dehydrogenase (LDH) and thereby led to inaccurate results in diagnostic assays. Beillmann et al [27] characterised one as an adduct of phosphate and NAD while Gallati [26] found that the identity of the buffer played an important role in the degree of LDH inhibition.

Our interest in NAD relates to the development of enzyme-based biosensor electrodes containing NAD-dependent dehydrogenases, such as glucose dehydrogenase (GDH) and D-3-hydroxybutyrate dehydrogenase (HBDH), and a redox mediator, such as 1,10-phenanthroline-5,6-dione, capable of oxidising NADH [29]. Here, the dry test strips must maintain their response for a minimum of 18 months. Stabilisation of the NAD cofactor together with the enzyme and mediator in the electrode are important in achieving this target. In this article, we report the development of a validated HPLC assay for NAD and apply it to stability studies of NAD in buffered solution and dry test strips.

2. EXPERIMENTAL

2.1 Materials

Methanol (HPLC-grade), potassium dihydrogen phosphate (KH_2PO_4), and sodium hydroxide (NaOH) were obtained from VWR International (Poole, UK). Buffer salts and ADP-ribose were purchased from Sigma-Aldrich chemicals (Gillingham, UK). The sodium salt of NAD was obtained from Oriental Yeast Co. (Tokyo, Japan). Nicotinamide, adenosine, adenine, AMP, and ADP were purchased from Alfa Aesar (Heysham, UK). De-ionised distilled water was used throughout the experiment. All other reagents were of analytical grade.

The dry test strips containing NAD and GDH were Optium Plus™ glucose electrodes from Abbott Diabetes Care (Maidenhead, UK). The test strips were supplied packaged in foil at low humidity.

2.2 HPLC Instrumentation and Conditions

A PerkinElmer (Norwalk, CT) HPLC system equipped with a module series 200 UV-vis detector, series 200 LC pump, series 200 autosampler and series 200 peltier LC column oven were used in this work. The data were acquired via PE TurboChrom Workstation data acquisition software using PE Nelson series 600 LINK interfaces. Chromatographic separation was achieved with Xterra RP 18, 5 μm (150 × 4.6 mm) which has a stationary phase based on silica gel where methyl groups are incorporated to reduce the number of surface silanol groups free to interact with basic compounds during chromatography. The column temperature was set at 35 °C and UV detection is set in the range of 200-400 nm. Mobile phase consisting of 20 mM potassium dihydrogen orthophosphate pH 8.0-methanol (95:5, v/v) was used at a flow rate of 1.0 mL/min. Injection volume was 10 μL. The mobile phase was filtered through 0.45 μm membrane filter and continuously degassed on-line. The second HPLC system used to measure the UV spectra was consisted of a Waters Alliance 2690 Separations Module to a 996 Waters photodiode array (PDA) Detector (Waters, Elstree, UK).

2.3 Preparation of Solutions and Samples

2.3.1 Solution preparation

An accurately weighed amount (30 mg) of NAD was placed in a 100 mL volumetric flask and dissolved in the mobile phase to produce a standard solution. Aliquots of this solution were diluted with the mobile phase to produce solutions with NAD concentrations in the range 0.01-0.7 mg/mL.

For solution stability studies, a 10mM solution of NAD (0.686 g) in the relevant buffer solution (100 mL, 100 mM, pH 7.0 and 8.0) was prepared.

2.3.2 Preparation of dry test strips

For solid-state stability studies, dry test strips (Optium Plus™) were stored as supplied in foil packets at temperatures of 4, 30, 40 and 50°C. At the relevant time point, ten test strips for each storage temperature were removed from their foil packets and prepared for NAD extraction. This involved peeling off a covering tape on each test strip then cutting off the end of each strip containing the dried active reagents including NAD. The ten strip pieces were added to a test tube containing 1.0 mL of water then vortex-mixed for 40s. The extract was finally transferred to a vial for HPLC analysis.

3. RESULTS AND DISCUSSION

3.1 Chromatographic Separation

The chromatographic analysis of NAD was carried out in the isocratic mode using a mixture of phosphate buffer-methanol (95:5, v/v) as mobile phase. The column was equilibrated with the mobile phase flowing at 1.0 mL/min for 30 min prior to injection. 10 μL of sample solutions were injected automatically into the column. The optimal wavelength for NAD detection was established using two UV absorbance scans over the range of 200 to 400 nm, one scan of the mobile phase, and the second of the analyte in the mobile phase. As expected, 260 nm was shown to be the optimal wavelength to maximise the signal (Figure 1). Chromatograms of the NAD gave good peak shape. The retention time for NAD was 2.0 min. To evaluate the quantitative nature of the analytical method, a series of samples with different amounts of NAD were run to investigate the best assay concentration. Using a C_{18} column, best concentration was assessed by injecting six reference standard solutions of NAD in the range of 0.01-0.7 mg/mL. The integrated peak areas were plotted versus amount injected. The calibration curve was found to be linear from concentration range 0.1-0.5 mg/mL with a correlation coefficient of 1.000. On the bases of these data, 0.3 mg/mL was chosen as a working concentration for the assay.

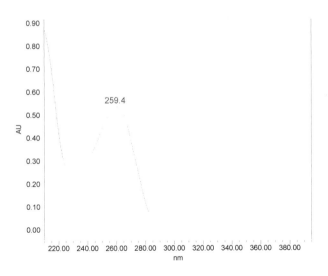

FIGURE 1 UV spectrum of a standard solution of NAD measured by HPLC-PDA.

System suitability testing was performed to determine the accuracy and precision of the system from six replicate injections of a solution containing 0.30 mg NAD/mL. The percent relative standard deviation (%RSD) of the retention time (min) and peak area were found to be less than 0.40%. The retention factor (also called capacity factor, k) was calculated using the equation $k = (t_r / t_0) -1$, where t_r is the retention time of the analyte and t_0 is the retention time of an unretained compound; in this study, t_0 was calculated from the first disturbance of the baseline after injection and capacity factor value was obtained 8.38 for NAD peak. The separation factor (α) was calculated using the equation, $\alpha = k_2 / k_1$ where k_1 and k_2 are the retention factors for the first and last eluted peaks respectively. The separation factor for NAD peak was 2.16 obtained. The plate number (also known as column efficiency, N) was calculated as $N = 5.54 (t_r / w_{0.5})^2$ where $w_{0.5}$ is the peak width at half peak height. In this study, the theoretical plate number was 2674. Resolution is calculated from the equation $R_s = 2(t_2 - t_1) / (t_{w1} + t_{w2})$. Where t_1 and t_2 are retention times of the first and second eluted peas, respectively, and t_{w1} and t_{w2} are the peak widths. The resolution for NAD peak was > 2.0. The asymmetry factor (A_s) was calculated using the US Pharmacopeia (USP) method. The peak asymmetry value for each NAD peak was 1.02.

Robustness studies were also performed in the method development phase applying the experimental design as shown in Table 1. A sample of NAD was prepared at working concentration (0.3 mg/mL) and assayed using the experimental design with eight test combinations for seven different chromatographic parameters as shown in Table 2. For each parameter, four combinations of (AAAA) and four combinations of (aaaa) were studied. The actual value of each parameter (V_A - V_G) (Table 2) shows which parameter has a dominant influence on the developed analytical method. In all cases good separations of NAD was always achieved, indicating that the analytical method remained selective for NAD component under the optimized conditions.

TABLE 1 Experimental design for robustness study

Test parameter	1	2	3	4	5	6	7	8
A/a	A	A	A	A	a	a	a	a
B/b	B	B	b	b	B	B	b	b
C/c	C	c	c	c	C	c	C	c
D/d	D	D	d	d	d	d	D	D
E/e	E	e	E	e	e	E	e	E
F/f	F	f	f	F	F	f	f	F
G/g	G	g	g	G	g	G	G	g
Results	s	t	u	v	w	x	y	z

TABLE 2 Chromatographic parameter for robustness study

Parameter	Test conditions 1	Test conditions 2	Differences
Analytical column	A = column C-18	a = column C-18	$V_A = ¼ (s+t+u+v) − ¼ (w+x+y+z) = A − a$
Sample solvent	B = Buffer/water	b = Mobile phase	$V_B = ¼ (s+t+w+x) − ¼ (u+v+y+z) = B − b$
Temperature	C = 30 °C	c = 40 °C	$V_C = ¼ (s+u+w+y) − ¼ (t+v+x+z) = C − c$
Flow rate	D = 0.8 mL/min	d = 1.3 mL/min	$V_D = ¼ (s+t+y+z) − ¼ (u+v+w+x) = D − d$
Wavelength	E = 250 nm	e = 270 nm	$V_E = ¼ (s+u+x+z) − ¼ (t+v+w+y) = E − e$
Mobile phase	F = 3% methanol	f = 7 % methanol	$V_F = ¼ (s+v+w+z) − ¼ (t+u+x+y) = F = f$
Solubility stability	G = 1 h	g = 8 h	$V_G = ¼ (s+v+x+y) − ¼ (t+u+w+z) = G - g$

3.2 Method Validation

According to best practice [30-32], the newly developed analytical method was validated in terms of linearity, assay range, precision (repeatability and intermediate precision), accuracy (recovery), specificity, limits of detection and quantification.

3.2.1 Linearity and range

The linearity of peak area response versus concentration for NAD was studied from approximately 0.1 to 0.5 mg/mL. Five solutions were prepared corresponding to 10, 25, 75, 100, and 150% of the nominal analytical concentration (0.3 mg/mL) and each one was injected in triplicate. The calibration curve was linear in the range of 0.1 to 0.5 mg/mL for this assay, with a correlation coefficient (R^2) for NAD (Table 3). A typical calibration curve has the regression equation of $y = 12058264.90x + 67726.33$.

3.2.2 Precision (repeatability and intermediate precision)

The precision of the analytical method was evaluated in terms of repeatability and intermediate precision. Repeatability was assessed on the same day by means of repetitive application of the NAD procedure to two samples (experimental replicates) at each concentration level (0.1, 0.2, 0.3, 0.4 and 0.5 mg/mL), and each one was injected in triplicate (instrumental replicates). Intermediate precision was assessed on six replicate injections at three different concentrations (0.1, 0.3 and 0.5 mg/mL) for two consecutive days. The precision data and RSD values presented in Table 3 were less than 0.40% in all cases and illustrated excellent precision for the analytical method.

TABLE 3 Validation results obtained for the HPLC assay of NAD

Validation criterion	Concentration range (mg/mL)	Results
Linearity	0.10 to 0.50	$y = 12058264.90x + 67726.33$
($n = 3$; $k = 5$)		$R^2 = 1.000$
Precision		
(A) Intra-day (%RSD; $n = 6$)	0.1, 0.2, 0.3, 0.4, 0.5	0.21, 0.32, 0.20, 0.27, 0.35
(B) Inter-day		
(%RSD; 2 days; $n = 6$)	0.1, 0.3, 0.5	0.26, 0.22, 0.38
LOD		(s/n = 3.3), 8.8 µg/mL
LOQ		(s/n = 10.0), 26 µg/mL

3.2.3 Accuracy/recovery study

In order to test the efficiency of the analytical method, recovery studies at three known added concentration levels (0.2, 0.3 and 0.4 mg/mL, 75%, 100% and 150% of nominal) were carried out. Three replicates were injected at each concentration level. Mean recoveries higher than 97.5% were obtained in all cases with RSD less than 0.17%. The results are shown in Table 4.

TABLE 4 Recovery studies of NAD from samples with known concentration

Conc (mg/mL)	Percent of nominal	Amount of NAD (mg)		Recovery (%)	RSD (%)	t-test
		Added	Recovered			
0.2	75	0.102	0.097	97.46	0.012	0.041
0.3	100	0.301	0.302	100.78	0.003	0.071
0.4	150	0.501	0.494	98.87	0.161	0.024
Mean				99.04		

3.2.4 Limits of detection and quantitation

The limit of detection (LOD) was considered as the minimum analyte concentration yielding a signal-to-noise ratio equal to three. The limit of quantitation (LOQ) was adopted at the lowest analyte concentration yielding a signal 10 times greater than the noise. The LOD and LOQ values were for NAD were found to be 8.8 μg/mL (s/n = 3.3) and 26 μg/mL (s/n = 10.0), with RSD 0.88% for six injections respectively (Table 3).

3.2.5 Specificity

Forced degradation studies were performed to evaluate the specificity of NAD under four stress conditions (heat, UV light, acid, base). Solutions of NAD were exposed to 50°C for 1 h, UV light using a Mineralight UVGL-58 light for 24 h, acid (1M HCl) for 24 h and base (1M NaOH) for 4 h. A summary of the stress results is shown in Table 5. The method was capable of separating NAD from various degradation products such as AMP, ADP, AD-ribose, nicotinamide, adenine and adenosine.

TABLE 5 Specificity results of NAD under stress conditions

Stress conditions	Sample treatment	t_R (min)	Peak area (μVs)	Assay (%)
Reference	Fresh solution	2.048	3694298	95.77
Acid	1M HCl for 24 h	2.046	3705876	95.73
Base	1M NaOH for 4 h	2.059	3714595	95.82
Heat	50 °C for 1 h	2.062	3715813	95.82
Light	UV Light for 24 h	2.046	3711822	95.71

4. METHOD APPLICATION

The validated HPLC method was applied to the determination of the stability of NAD under various conditions. Aqueous solutions of NAD containing a wide range of different buffers were investigated at neutral and alkaline pH. These included inorganic anions and zwitterionic aminosulfonic acids, the so-called Good buffers [33-35]. In a second application of the HPLC method, the stability of NAD in a dry test strip incorporating single selected buffer salt was assessed at temperatures in the range 4-50°C.

5. STABILITY OF NAD IN AQUEOUS BUFFERED SOLUTION

A range of buffers with different structural features was evaluated in terms of the amount of NAD remaining after heating at 50°C for 24 h (Table 6). The buffered aqueous solutions of NAD were assessed against a control comprising a simple solution of NAD in water, i.e., no buffer. In general, the zwitterionic Good buffers tended to have little effect on the stability of NAD in solution at pH 7. The level of residual NAD varied over a small range from 66.8% for MES to 70.4% for CAPS compared to 69.2% for the water control. However, there was a small but definite trend of increasing residual NAD with increasing buffer pK_a for these zwitterionic buffers. This is believed to be simply due to greater protonation of the amine group in the buffers with high pK_a at pH 7, leading to less nucleophilic attack on NAD. In practice, one would not use the high pK_a buffers such as CAPS and TAPS at pH 7 since this lies outside their effective buffering range.

The correlation between buffer pK_a and residual NAD breaks down when one attempts to include buffers of a different structural class to the zwitterionic Good ones. Thus, the 1° amine TRIS has a relatively high residual NAD level of 69.9% for its pK_a while the 2° diamine bis-TRIS-propane gives a low level of 59.6%. The inorganic anions, carbonate and borate, are clearly very reactive towards NAD yielding very low residual NAD results of 2.2% and 26.2% respectively.

It is interesting to note that the profile of NAD decomposition products depends on the buffer. Both BES and borate gave almost exclusively ADP-ribose and nicotinamide, i.e., the first step in the decomposition of NAD [17], when solutions at pH 8.0 were heated at 60°C for 24 h (Figure 2a & c). In contrast, carbonate and TRIS also gave AMP and ADP, presumably derived from the decomposition of ADP-ribose [17, 18]; especially high levels of these two nucleotides were seen in the former case (Figure 2b & d). In addition, the NAD decomposition profile in TRIS displayed an unidentified shoulder on the ADP-ribose peak.

6. STABILITY OF NAD IN DRY TEST STRIPS

As a result of the solution studies in Section (stability of NAD in aqueous buffered solution), BES was selected as the buffer for use in a dry test strip for blood glucose (Optium Plus™) containing the enzyme GDH and NAD. A high proportion of NAD was found to be retained in BES buffer solutions at elevated temperatures while any decomposition of NAD resulted in the clean formation of nicotinamide and ADP-ribose, species that were not anticipated to interfere with the functioning of the test strip. The active reagents including NAD on the test strip were buffered at pH 7.0 close to physiological pH which is within the effective buffering range of BES (Table 6).

Five lots of dry glucose test strips (Optium Plus™), containing NAD and BES buffer, were heated at 4°C, 30°C, 40°C and 50°C for three months. NAD was readily extracted from the dry strip into water and then analysed by HPLC using the validated method. The stability of NAD in test strips at 30°C, 40°C and 50°C was assessed by determining the residual NAD at each temperature versus control test strips stored at 4°C. Mean residual NAD levels for the five test strip lots were 96.6%, 92.8% and 82.6% at 30°C, 40°C and 50°C respectively (Table 7, Figure 3). Nicotinamide and ADP-ribose were the only detected decomposition products of NAD. These results show that NAD is very stable in dry test strips (Optium Plus™) packaged in foil under low humidity and stored at temperatures in the range 30 - 50°C. Currently, the test strips are marketed successfully with a shelf life of 18 months at a maximum storage temperature of 30°C.

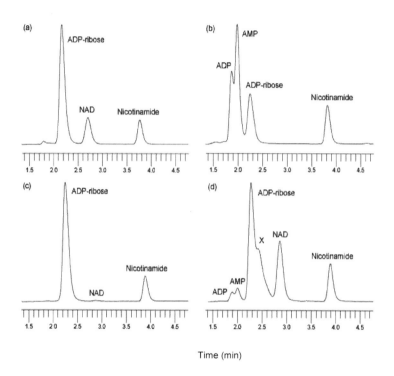

FIGURE 2 Chromatograms of NAD solutions heated at 60°C for 24 h in the presence of various buffer salts at pH 8.0: (a) 0.1 M BES, (b) 0.1 M carbonate, (c) 0.1 M borate, (d) 0.1 M TRIS where X is an unidentified species.

TABLE 6 Stability of 10 mM NAD solutions in 100 mM buffer (or additive) pH 7.0: residual NAD by HPLC after heating at 50°C for 24 h

Buffer / additive	Structure	Buffer pKa (effective pH range)	Residual NAD (%)
Water control	-	-	69.2
CAPS		10.4 (9.7 – 11.1)	70.4
TAPS		8.4 (7.7 – 9.1)	70.1
Bicine		8.3 (7.6 – 9.0)	68.7
TRIS		8.1 (7.0 – 9.0)	69.9
Tricine		8.1 (7.4 – 8.8)	68.6
EPPS (or HEPPS)		8.0 (7.3 – 8.7)	68.4
TES		7.5 (6.8 – 8.2)	68.4
HEPES		7.5 (6.8 – 8.2)	67.9
MOPS		7.2 (6.5 – 7.9)	67.5
BES		7.1 (6.4 - 7.8)	67.5
PIPES		6.8 (6.1 – 7.5)	67.7
MES		6.1 (5.5 - 6.7)	66.8
Bis-TRIS-propane		6.8, 9.0 (6.3 – 9.5)	59.6
Carbonate	-	6.4	2.2
N-Methyl-D-glucamine		-	50.8
Borate	-	9.1, 12.7, 13.8	26.2

TABLE 7 Stability of NAD in five lots of glucose test strips: residual NAD by HPLC after storage at 30°C, 40°C and 50°C for 3 months

Test strip lot no.	Residual NAD (%) versus 4°C		
	30°C	40°C	50°C
1	96.9	94.4	85.1
2	100.3	93.4	82.1
3	95.0	93.4	82.9
4	94.5	91.1	83.3
5	96.1	91.5	79.8
Mean:	96.6	92.8	82.6

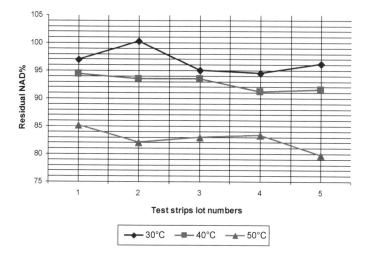

FIGURE 3 Plot for stability of NAD in five lots of glucose test strips: residual NAD by HPLC after storage at 30°C, 40°C and 50°C for 3 months.

7. CONCLUSION

A new, simple and robust reversed-phase HPLC method for the determination of NAD and the identification of its decomposition products in solution and solid-phase stability studies was developed and validated. Experiments in solution in the presence of a number of zwitterionic Good buffers demonstrated that there was a small trend of increasing residual NAD with increasing buffer pK_a. The decomposition profile of NAD in solution depended on the identity of the buffer. Following these studies, BES was selected as an ideal buffer for incorporation into dry test strips to maintain a high residual NAD level on storage while also restricting the products of any decomposition to nicotinamide and ADP-ribose as demonstrated by HPLC analysis. The resulting commercial test strips (Optium Plus™) for glucose have a long shelf life.

REFERENCES

1. Futatsugi, M; Oonogi, H; Date, M; Mimata, Y; Yamaguchi, F. Stable nicotinamide adenine dinucleotide derivative for use in dehydrogenase assays. WO 2001,094,370 (13 Dec 2001).
2. Dorn, A. R.; Hurt, C. J.; Ganser, E. O. Stable aqueous reagent containing NAD for alcohol and lactate determination. US Patent 5,804,403 (08 Sep 1998).
3. Takayama, S.; Deguchi, T.; Fujiki, J. Stabilizers for creatine kinase assay reagents and the stabilized aqueous reagents. Japanese Patent 10,327,895 (15 Dec 1998).
4. Nagel, R.; Mistele, J.; Schroder, N. Creatine kinase determination from blood using a reducing sulfur compound stabilized assay reagent in a spectrophotometric multistep enzyme assay. WO 1999,029,895 (17 Jun 1999).
5. Zielenski, R.; Nagel, R. Stabilized coenzyme solutions for the determination of dehydrogenases or their substrates. German Patent 10,001,529 (19 Jul 2001).

6. Zielenski, R. Stabilized coenzyme solutions and their use thereof for the determination of dehydrogenases or the substrate thereof in an alkaline medium. US Patent 6,162,615 (19 Dec 2001).

7. Nagai, T. Stable injection of nicotinamide adenine dinucleotide. Japanese Patent 45,012,153 (1970).

8. Staerk, J. R. Mueller-Matthesius, R. Stabilized nicotinamide-nucleotides. German Patent 2,814,154 (11 Oct 1979).

9. Bueckmann, A. F.; Kula, M. R.; Wichmann, R.; Wandrey, C. An efficient synthesis of high-molecular-weight NAD(H) derivatives suitable for continuous operation with coenzyme-dependent enzyme systems. *J. Appl. Biochem.* 1981, *3*, 301-315.

10. Adachi, S.; Ogata, M.; Tobita, H.; Hashimoto, K. Effects of molecular weight of dextran and NAD+ density on coenzyme activity of high molecular weight NAD+ derivative covalently bound to dextran. *Enzyme Microb. Technol.* 1984, *6*, 259-262.

11. Colombo, B. M.; Causa, P.; Primavera, P.; Villano, V. Stability predictions for pharmaceutical preparations. Il Farmaco Ed. Pr. 1972, *27*, 327-333.

12. Stocchi, V.; Cucchiarini, L.; Canestrari, F.; Piacentini, M. P.; Fornaini, G. A very fast ion-pair reversed-phase HPLC method for the separation of the most significant nucleotides and their degradation products in human red blood cells. *Anal. Biochem.* 1987, *167*, 181-190.

13. Crescentini, G.; Stocchi, V. J. Fast reversed-phase high-performance liquid chromatographic determination of nucleotides in red blood cells. *Chromatogr.* 1984, *290*, 393-399.

14. Stocchi, V.; Cucchiarini, L.; Magnani, M.; Chiarantini, L.; Palma, P.; Crescentini, G. Simultaneous extraction and reverse-phase high-performance liquid chromatographic determination of adenine and pyridine nucleotides in human red blood cells.*Anal. Biochem.* 1985, *146*, 118-124.

15. Anderson, F. S.; Murphy, R. C. Isocratic separation of some purine nucleotide, nucleoside, and base metabolites from biological extracts by high-performance liquid chromatography. *J.Chromatogr.* 1976, *121*, 251-262.

16. Chenault, H. K.; Whitesides, G. M. Regeneration of nicotinamide cofactors for use in organic synthesis. Appl. Biochem. Biotechnol. 1987, *14*, 147-197.

17. Kaplan, N. O.; Colowick, S. P.; Barnes, C. C. The effect of alkali on diphospho pyridine nucleotide. *J. Biol. Chem.* 1951, *191*, 461-472.

18. Schlenk, F.; von Euler, H.; Heiwinkel, H.; Gleim, W.; Nyström, H. The action of alkali on cozymase. Z. *Physiol. Chem.* 1937, *247*, 23-33.

19. (a) Pfleiderer, G.; Sann, E.; Stock, A. The mechanism of action of dehydrogenases. The reactivity of pyridine nucleotides (PN) and PN-models with sulfite as the nucleophilic reagent. Chem Ber. 1960, *93*, 3083-3099. (b) Johnson, S. L.; Smith, K. W. The interaction of borate and sulfite with pyridine nucleotides. Biochemistry, 1976, *15*, 553- 559.

20. Yarmolinsky, M. B.; Colowick, S. P. Mechanism of pyridine nucleotide reduction by dithionite. Biochim. Biophys. Acta. 1956, *20*, 177-189.

21. Colowick, S. P.; Kaplan, N. O.; Ciotti, M. M. The reaction of pyridine nucleotide with cyanide and its analytical use. *J. Biol. Chem.* 1951, *191*, 447-459.

22. Eys, J. V.; Kaplan, N. O. Addition of sulfhydryl compounds to diphosphopyridine nucleotide and its analogs. *J. Biol. Chem.* 1957, *228*, 305-314.

23. Burton, R. M.; Kaplan, N. O. The reaction of pyridine nucleotides with carbonyl compounds. J. Biol. Chem. 1954, *206*, 283-297.

24. Burton, R. M.; Kaplan, N. O. A chemical reaction of hydroxylamine with diphosphopyridine nucleotide. *J. Biol. Chem.* 1954, *211*, 447-463.

25. Anderson, B. M.; Anderson, C. D. The effect of buffers on nicotinamide adenine dinucleotide hydrolysis. *J. Biol Chem.* 1963, *238*, 1475-1478.

26. Gallati, H. Formation and purification of the oxidoreductase inhibitor from NAD. *J. Clin. Chem. Clin. Biochem.* 1976, *14*, 3-8.

27. Biellmann, J-. F.; Lapinte, C.; Haid, E.; Weimann, G. Structure of lactate dehydrogenase inhibitor generated from coenzyme. *Biochemistry.* 1979, *18(7)*, 1212-1217.

28. Godfredsen, S. E.; Ottesen, M.; Andersen, N. R. On the mode of formation of 1,6-dihydro-NAD in NADH preparations. *Carlsberg Res. Commun.* 1979, *44*, 65-75.

29. Forrow, N. J.; Sanghera, G. S.; Walters, S. J.; Watkin, J. L. Development of a commercial amperometric biosensor electrode for the ketone D-3-hydroxybutyrate. Biosens. *Bioelectron.* 2005, *20(8)*, 1617-1625.

30. Shabir, G.A.; Lough, J.W.; Shafique, A.A.; Bradshaw, T.K. Evaluation and Application of Best Practice in Analytical Method Validation. *J. Liq. Chromatogr. Rel. Technol.* 2007, *30(3)*, 311-333.

31. Shabir, G.A. Validation of high-performance liquid chromatography methods for pharmaceutical analysis: Understanding the differences and similarities between validation requirements of the US Food and Drug Administration, the US Pharmacopeia and the International Conference on Harmonization. *J. Chromatogr. A.* 2003, *987(1-2)*, 57-66.

32. Shabir, G. A. Step-by-step analytical methods validation and protocol in the quality system compliance industry. *J. Validation Technol.* 2004, *10*(4), 314-324.

33. Good, N. E. Uncoupling of the Hill reaction from photophosphorylation by anions. *Arch. Biochem. Biophys.* 1962, *96*, 653-661.

34. Good, N.E.; Winget, G.D.; Winter, W.; Connolly, T.N.; Izawa, S.; Singh, R.M.M. Hydrogen ion buffers for biological research. *Biochem.* 1966, *5*, 467-477.

35. Ferguson, W. J.; Braunschweiger, K. I.; Braunschweiger, W. R.; Smith, J. R.; McComick, J. J.; Wasmann, C. C.; Jarvis, N. P.; Bell, D. H.; Good, N. E. Hydrogen ion buffers for biological research. *Anal. Biochem.* 1980, *104(2)*, 300-310.

24 A NEW VALIDATED METHOD FOR THE SIMULTANEOUS DETERMINATION OF A SERIES OF EIGHT BARBITURATES USING HPLC

ABSTRACT

A new reversed-phase high performance liquid chromatographic (RP-HPLC) method is developed and validated for the simultaneous determination of barbitone, allobarbitone, phenobarbitone, cyclobarbitone, hexobarbitone, pentobarbitone, secobarbitone and methohexitone compounds in a single analytical run. The method uses a Phenosphere C_{18} (150 mm × 4.6 mm; 5 μm) column and isocratic elution. The mobile phase consisted of a mixture of methanol-water (50:50, v/v), pumped at a flow rate of 1.0 mL/min. The UV detection is set at 254 nm. The method is validated with respect to accuracy, precision (repeatability and intermediate precision), specificity, linearity, range robustness and stability of analytical solutions. All the parameters examined met the current recommendations for bioanalytical method validation. The method is specific, simple, selective and reliable for routine use in quality control analysis of barbiturates raw materials for final product release.

Keywords: Barbitone, Allobarbitone, Phenobarbitone, Cyclobarbitone, Hexobarbitone, Pentobarbitone, Secobarbitone, Methohexitone, HPLC, Method development; Method validation

1. INTRODUCTION

Barbiturates are widely in use since the beginning of the century (barbital, 1903) especially as sedative hypnotics. With the advent of anxiolytic agents the popularity of barbiturates has suffered although they are still less costly. However, those with specialized properties such as the anticonvulsant phenobarbital continue to be commonly used [1]. In addition, abuse of barbiturates is now widespread. Due to the international nature of the illegal drug market forensic laboratories encounter a vast range of such compounds. Complications arise from the fact that abused barbiturates often occur as complex mixtures and other drugs and / or excipients are also present [2]. This necessitates the continued development of methods for their efficient separation and precise identification. In this work a group of eight barbiturates was selected as model mixtures (Figure 1). Barbitone (BR) is also known as 5,5-diethylbarbituric acid, colours crystals or white crystalline powder with melting point 188-192°C. It is seldom used in modern therapeutics. Allobabitone (AB) known as 5,5-diallybarbituric acid, a white crystalline powder, melting point about 173°C, dissociation constant at Pka 7.8 at 25°C. Phenobarbitone (PhB) is also known as 5-ethyl-5-phenylbarbituric acid, white crystalline powder, and melting point 174-178°C, Pka 7.4 at 25°C. Cyclobarbitone (CB) is known as

5-(cyclohex-1-enyl)-5-ethylbarbituric acid, melting point 171-175°C, Pka 7.6 at 20°C. Hexobarbitone (HB) is known as 5-(cyclohex-1-enyl)-1,5-dimethyl barbituric acid, melting point 144-148°C, Pka 8.2 at 20°C. Pentobarbitone (PB) is known as 5-(ethyl-5(1-methylbutyl) barbituric acid, melting point from 127-133°C. Secobarbitone (SB) is known as 5-allyl-5-(1-methylbutyl) barbituric acid, melting point about 100°C, very slightly soluble in water, freely soluble in ethanol and ether, also soluble in chloroform, Pka 7.9 at 20°C. Methohexitone (MH) is known as $\propto$-(±)-5-allyl-1-methyl-5-(methypent-2-ynyl) barbituric acid. A white to faintly yellowish-white crystalline powder with melting point 92-96°C.

Some methods for the determination of some barbiturates have been reported, such as micellar liquid chromatography [3,4], supercritical fluid chromatography [5], gas chromatography-mass spectrometry (GC/MS) [6], thin-layer chromatography (TLC) [7, 8], high-performance thin-layer chromatography (HPTLC) [9], Capillary Electrophoresis (CE), [10] high-performance liquid chromatography (HPLC) [11-14], and gas chromatography (GC) [15, 16], but simultaneous determination of a series of eight barbiturate by reversed-phase HPLC and method validation has not been reported.

Furthermore, most of these procedures reported require labour sample pre-treatment and solvent extraction or solid phase extraction. These pre-treatment steps are time-consuming, increase the error sources and make the procedure more laborious.

Analytical methods validation is an important regulatory requirement in pharmaceutical analysis. In recent years, the International Conferences on Harmonization (ICH) has introduced guidelines for analytical methods validation [17] in Japan Europe and United States. The most widely applied analytical performance characteristics are accuracy, specificity, linearity, range, precision (repeatability and intermediate precision), stability of analytical solutions and robustness.

The purpose of this study was to develop and validate a rapid, accurate, simple and robust reversed-phase HPLC method for the simultaneous determination of a series of eight barbiturates (barbitone, allobarbitone, phenobarbitone, cyclobarbitone, hexobarbitone, pentobarbitone, secobarbitone and methohexitone) in a single chromatographic run which can be reliably used in routine quality control analysis for final drug substances release for medicinal formulations.

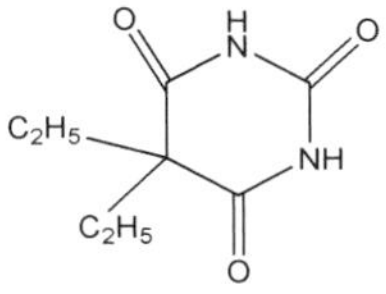

FIGURE 1 Chemical structures of the barbiturates used in this study.

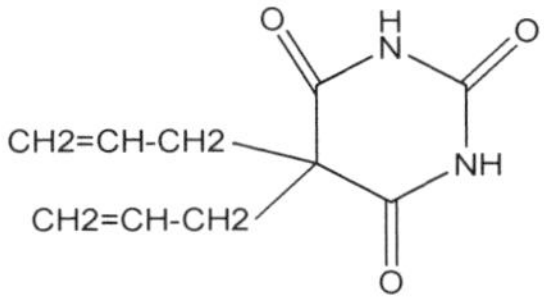

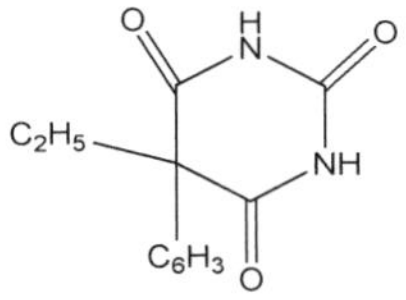

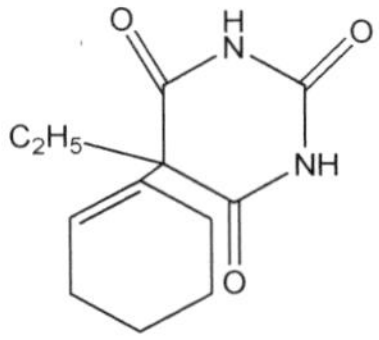

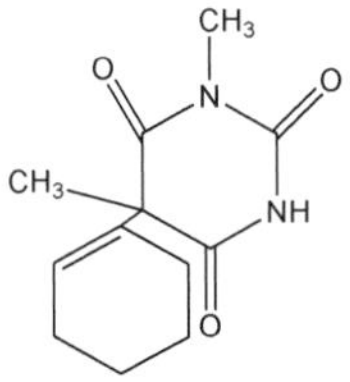

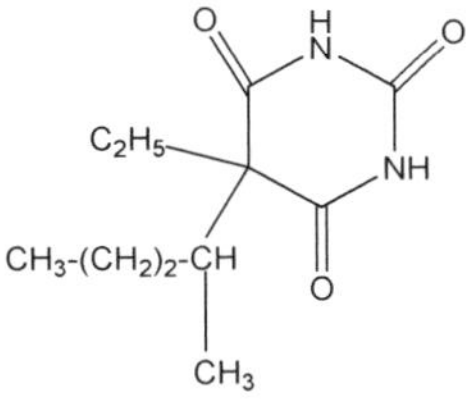

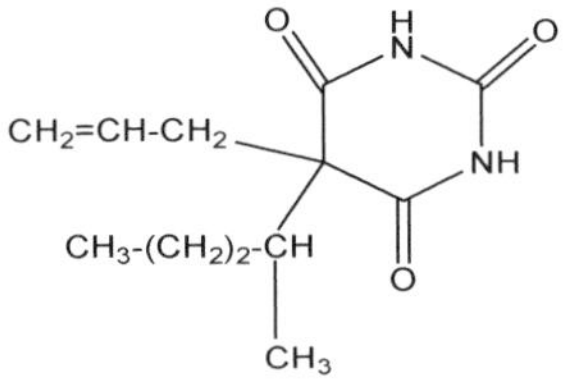

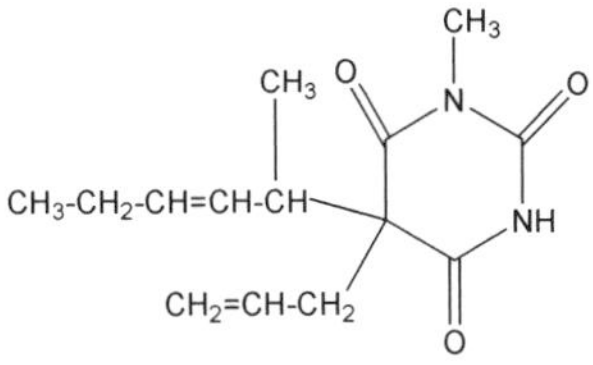

2. EXPERIMENTAL

2.1 Materials

All chemicals and reagents were of the highest purity. Methanol
(HPLC-grade) and barbitone, allobarbitone, phenobarbitone,
cyclobarbitone, hexobarbitone, pentobarbitone, secobarbitone and
methohexitone were purchased from Sigma (Gillingham, UK).
Distilled water was de-ionised by using a Milli-Q system (Millipore,
Bedford, MA).

2.2 HPLC Instrumentation and Conditions

A Knauer (Berlin, Germany) HPLC system equipped with a module
1000 LC pump, LC 3950 autosampler, LC 2600 photodiode-array
(PDA) detector and a vacuum degasser. The data were acquired
via Knauer ClarityChrom workstation data acquisition software. All
chromatographic experiments were performed in the isocratic mode.
The mobile phase consisted of a mixture of methanol-water (50:50,
v/v). The flow rate was set to 1.0 mL/min. The injection volume was
10 µL and the detection wavelength was set at 254 nm. The
chromatographic separation was carried out on a 150 mm × 4.6 mm,
5 µm C_{18} Phenosphere column obtained from Phenomenex
(Macclesfield, UK).

2.3 Sample Preparation

An accurately weighted amount (0.08 g) of barbitone, (0.039 g) allobarbitone, (0.037 g) phenobarbitone, (0.07 g) cyclobarbitone, (0.048 g) hexobarbitone, (0.070 g) pentobarbitone, (0.076 g) secobarbitone and (0.07 g) methohexitone were placed in a 100 mL volumetric flask and dissolved in mobile phase (stock). A 5 mL aliquot of stock solution was diluted to 100 mL volumetric flask in mobile phase, yielding a final concentration of 400, 195, 185, 35, 24, 35, 38, and 35 µg/mL, respectively.

3. RESULTS AND DISCUSSION

3.1 Method Development

The chromatographic separation of barbitone, allobarbitone, phenobarbitone, cyclobarbitone, hexobarbitone, pentobarbitone, secobarbitone and methohexitone was carried out in the isocratic mode using a mixture of methanol-water (50:50, v/v) as mobile phase. The column was equilibrated with the mobile phase flowing at 1.0 mL/min for about 30 min prior to injection. The column temperature was ambient. 10 mL of standard solutions was injected automatically into the column. Subsequently, the liquid chromatographic behaviours of barbiturates were monitored with a PDA UV detector at 254 nm. Additionally, preliminary system suitability, precision, linearity, robustness and stability of solutions studies performed during the development of the method showed that the 10 µL injection volume was reproducible and the peak response was significant at the analytical concentration chosen. Chromatograms of the resulting solutions gave good separation and resolution (Figure 2). The analysis time for standards for all compounds was ca.30 min.

System suitability test was developed for the routine application of the assay method. Prior to each analysis, the chromatographic system must satisfy suitability test requirements (resolution and repeatability). Peak-to-peak resolution, between each peak measured on a reference solution must be above 2. System suitability test was performed to determine the accuracy and precision of the system from six replicate injections of a solution containing 30 µg barbiturates / mL. All peaks were well resolved and the precision of injections for all preservative peaks were acceptable. The percent relative standard deviation (RSD) of the peaks area responses were measured, giving an average between 0.9% and 0.36% (n = 6). The tailing factor (T), capacity factor (K), theoretical plate number (N) and height equivalent to a theoretical plate (HETP) were also calculated. The results of system suitability in comparison with the required limits are shown in Table 1. The proposed method met these requirements within the accepted limits [18, 19].

For the determination of method robustness within a laboratory during method development a number of chromatographic parameters were evaluated, such as flow rate, column temperature, mobile phase composition, columns from different batches, and the quantitative influence of the variables were determined. For each

444

parameter studied two injections of standard solutions were chromatographed. In all cases the influence of the parameters were found within a previously specified tolerance range. This shows that the method for determination of barbitone, allobarbitone, phenobarbitone, cyclobarbitone, hexobarbitone, pentobarbitone, secobarbitone and methohexitone was reproducible and robust.

TABLE 1 System suitability test recommended limits and results of eight barbiturates

Parameters	Recommen ded limits	Results of barbiturates							
		1	2	3	4	5	6	7	8
Retention time (min)	-	2.82	4.11	4.65	6.68	8.83	11.52	15.58	24.63
Injection repeatability ($n = 6$)	R.S.D. ≤ 1 (%, $n \geq 5$)	0.09	0.14	0.11	0.25	0.16	0.19	0.16	0. 36
Resolution (Rs)	$R_s > 1.5$	-	2.56	2.06	4.06	4.30	5.38	8.12	18.10
Capacity factor (K')	> 2	2.65	2.87	3.37	5.27	7.30	9.83	13.64	22.14
Tailing factor (T)	≤ 2	0.625	0.750	0.750	1.000	0.887	0.875	0.875	1.000
HETP*	-	0.015	0.009	0.009	0.005	0.004	0.003	0.003	0.003
Theoretical plate (N)	>2000	2235	2491	2705	2852	3467	5023	5639	6737

*Height equivalent to a theoretical plate

3.2 Method Validation

3.2.1 Linearity and range

The linearity test was performed using five different amounts of barbitone, allobarbitone, phenobarbitone, cyclobarbitone, hexobarbitone, in the range 360 – 440, 155 – 230, 155 – 220, 5 – 65, 10 – 40 µg/mL, respectively and for pentobarbitone, secobarbitone and methohexitone 5 – 65 µg/mL. Solutions corresponding to each concentration level were injected in duplicate and linear regression analysis of the barbiturates peak area (y)

versus barbiturates concentration (x) was calculated. The correlation coefficients (r^2 = ≥ 0.9995) obtained for each barbiturate for the regression line demonstrates that there is a strong linear relationship between peak area and concentration of barbiturates (Table 1).

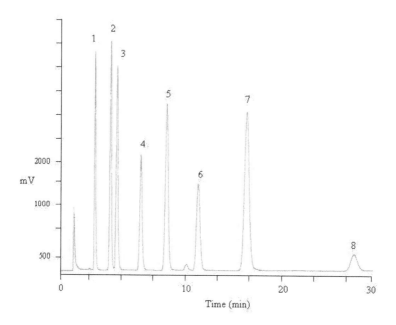

FIGURE 2 HPLC chromatogram of eight barbiturates: (1) barbitone, RT = 2.82, (2) allobarbitone, RT = 4.11, (3) phenobarbitone, RT = 4.65, (4) cyclobarbitone, RT = 6.68, (5) hexobarbitone, RT = 8.83, (6) pentobarbitone, RT = 11.52, (7) secobarbitone, RT = 15.58 and (8) methohexitone, RT = 24.63.

TABLE 2 Linearity assessment of the HPLC method for the assay of barbiturates

Components	Conc (μg/mL)[a]	Range (μg/mL)	Equation for regression line	r^2
Barbitone	400	360 - 440	y = 31.08x - 10591	0.9999
Allobarbitone	195	155 - 230	y = 33.118x - 4630.3	0.9996
Phenobarbitone	185	155 - 220	y = 44.533x - 6488.2	0.9998
Cyclobarbitone	35	5 - 65	y = 44.18x + 63.3	0.9995
Hexobarbitone	24	10 - 40	y = 80.876x - 370.58	0.9995
Pentobarbitone	35	5 - 65	y = 44.76x + 76.8	0.9997
Secobarbitone	38	5 - 65	y = 44.087x + 73.167	0.9998
Methohexitone	35	5 - 65	y = 44.687x + 57.967	0.9997

[a] Target concentration corresponding to 100%

3.2.2 Precision

The precision of the method was determined by repeatability (intra-day) and intermediate precision (inter-day variation). Repeatability was examined by analysing six determinations of the same batch of each component at 100% of the test concentration. The relative standard deviation (RSD) of the areas of barbiturates peak were found to be less than 0.36 % (Table 3), which confirms that the method is sufficiently precise.

Intermediate precision (inter-day variation) was studied by assaying five samples containing the nominal amount of barbiturates on different days. Solutions corresponding to each concentration level were injected in duplicate. The RSD values across the system were calculated and found to be less than 0.45% (Table 3) for each of the multiple sample preparation, which demonstrates excellent precision for the method.

TABLE 3 Method validation results for barbiturates

Validation steps	Parameters	Results							
		1	2	3	4	5	6	7	8
Repeatability	R.S.D. (%, n = 6)	0.22	0.13	0.19	0.35	0.20	0.11	0.19	0.26
Int. precision									
Day 1	R.S.D. (%)	0.22	0.18	0.24	0.17	0.09	0.11	0.22	0.19
Day 2	R.S.D. (%)	0.27	0.22	0.18	0.13	0.14	0.39	0.27	0.45
Standard stability (24 h data)	Change in response factor (%)	0.10	0.09	0.13	0.15	0.11	0.14	0.12	0.13
System suitability	R.S.D. (%, n = 6)	0.18	0.18	0.23	0.16	0.18	0.32	0.22	0.29

3.2.3 Accuracy/recovery study

Recovery studies may be performed in a variety of ways depending on the composition and properties of the sample matrix. In the present study, three different solutions were prepared with a known added amount of pure barbiturate compounds to give a concentration range of 50-150% of that in a test preparation. These solutions were injected in triplicate and percent recoveries of response factor (area/concentration) were calculated (Table 4).

TABLE 4 Recovery studies of the HPLC method for the assay of barbiturates

Components	Applied concentration (% of target) (n = 3)		
	50 (%)	100 (%)	150 (%)
Barbitone	99.98 ($\pm$ 0.14) [a]	99.92 ($\pm$ 0.17)	99.88 ($\pm$ 0.27)
Allobarbitone	99.97 ($\pm$ 0.21)	100.00 ($\pm$ 0.32)	99.84 ($\pm$ 0.35)
Phenobarbitone	99.93 ($\pm$ 0.16)	99.92 ($\pm$ 0.27)	99.80 ($\pm$ 0.17)
Cyclobarbitone	99.92 ($\pm$ 0.17)	100.00 ($\pm$ 0.32)	100.00 ($\pm$ 0.15)
Hexobarbitone	99.82 ($\pm$ 0.37)	99.92 ($\pm$ 0.44)	99.85 ($\pm$ 0.17)
Pentobarbitone	99.98 ($\pm$ 0.23)	98.94 ($\pm$ 0.30)	99.72 ($\pm$ 0.25)
Secobarbitone	99.84 ($\pm$ 0.09)	99.97 ($\pm$ 0.27)	99.79 ($\pm$ 0.48)
Methohexitone	99.95 ($\pm$ 0.39)	100.00 ($\pm$ 0.21)	99.84 ($\pm$ 0.25)

[a] The coefficient of variation

3.2.4 Specificity and selectivity

Injections of the blank were performed to demonstrate the absence of interference with the elution of the barbitone, allobarbitone, phenobarbitone, cyclobarbitone, hexobarbitone, pentobarbitone, secobarbitone and methohexitone barbiturates. These results demonstrate (Figure 3) that there was no interference from the other compounds and, therefore, confirm the specificity of the method.

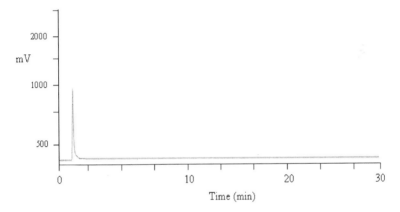

FIGURE 3 HPLC chromatogram of the blank run.

3.2.5 Stability of analytical solutions

Sample solutions were chromatographed immediately after preparation and then re-assayed after storage at room temperature for 24 h. The results given in Table 3 showed there was no significant change (< 0.16% response factor) in barbiturate concentrations over this period.

3.2.6 Measurement of robustness

Analytical methods developed for use in quality control laboratories ideally are robust. Retention time for the analytes of interest will not change significantly from day-to-day or from laboratory-to-laboratory if the method is considered robust. To determine the robustness of the chromatographic methodology developed for barbiturates, experimental conditions were purposely altered and chromatographic characteristics were evaluated. The effected temperature was also studied. Standard solutions were prepared and injected at early 20°C and again at 27°C. In all cases studied, the retention times of these compounds (barbitone, allobarbitone, phenobarbitone, cyclobarbitone, hexobarbitone, pentobarbitone, secobarbitone and methohexitone) were remains same 2.83, 4.12, 4.66, 6.68, 8.83, 11.51, 15.57 and 24.64 min, respectively (Figure 4). The coefficient of variation for retention time was lass then 1%. Good separation was always achieved, indicating that the analytical method remained selective for all components under the measured conditions.

3.2.7 System suitability

A system suitability test was performed to determine the accuracy and precision of the system by injecting six replicate injections of barbitone, allobarbitone, phenobarbitone, cyclobarbitone, hexobarbitone, pentobarbitone, secobarbitone and methohexitone standard solutions. The RSD of the peak areas responses was measured. The RSD for barbiturates was less then (0.33%) as can be seen in Table 3.

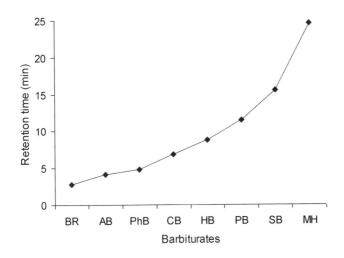

FIGURE 4 Retention times (min) of barbiturates: barbitone (BR), allobarbitone (AB), phenobarbitone (PhB), cyclobarbitone (CB), hexobarbitone (HB), pentobarbitone (PB), secobarbitone (SB), and methohexitone (MH).

4. CONCLUSION

A new RP-HPLC method with UV spectrophotometric detection was developed successfully for the simultaneous determination of a series of eight barbitone, allobarbitone, phenobarbitone, cyclobarbitone, hexobarbitone, pentobarbitone, secobarbitone and methohexitone compounds. The method was validated and the results obtained were accurate and precise with RSD $< 1\%$ in all cases and no significant interfering peaks were detected. The method is specific, simple, selective and reliable for routine use in quality control analysis of barbiturates raw materials for final product release.

REFERENCES

1. Gringauz, A. in: Introduction to Medicinal Chemistry, Wiley-VCH, 1997, p. 568-569.
2. Gill, R., Stead, A.H., Moffat, A.C. Analytical aspects of barbiturate abuse: Identification of drugs by the effective combination of gas-liquid, high-performance liquid and thin-layer chromatographic techniques. *J. Chromatogr. A* 1981, *204*, 275-284.
3. Marina, D., Rukhadze, G., Bezarashvili, S., Maya, V. S., Veronika, R. M. Separation of barbiturates with micellar liquid chromatography and optimization by a second order mathematical design. *J. Chromatogr. A* 1998, *805,* 45-53.
4. Martín, S., Sagrado, R. M., Villanueva, C. Medina, M. J. Determination of barbiturates in urine by micellar liquid chromatography and direct injection of sample *J. Pharm. Biomed. Anal.* 1999, *21*, 331-338.
5. Roger, M., Smith, M., Marsin, S. Application of packed column supercritical fluid chromatography to the analysis of barbiturates. *J. Pharm. Biomed. Anal.* 1988, *6*, 837-841.
6. Brad, J. H., Jennifer, S. B. Determination of barbiturates by solid-phase microextraction (SPME) and ion trap gas chromatography–mass spectrometry. *J. Chromatogra. A* 1997, *777*, 275-282.
7. Tibor, C., Jacek, B., Éva, F., Jozsef, S. Reversed-phase thin-layer chromatography of barbiturates in the presence of soluble β-cyclodextrin polymer. *J. Chromatogra. A* 1986, *351*, 356-362.
8. Grassini, G.S., Cristalli, M. Comparison of C_n bonded silica gel thin-layer chromatographic plates: conditions for use and separations of some barbiturates. *J. Chromatogra. A* 1981, *214, 209-216.*

9. Giuliana, G.S., Isabella, N. High-performance thin-layer chromatography on amino-bonded silica gel: application to barbiturates and steroids. *J. Chromatogra. A* 1985, *322*, 149-158.

10. Ting-Fu, J., Yuan, H. W., Zhi, L., Mei-E. Y. Direct Determination of Barbiturates in Urine by Capillary Electrophoresis Using a Capillary Coated Dynamically with Polycationic Polymers. *Chromatographia*, 2007, *65*, 611-615.

11. Fernández, G. B. P., Lores, M., Cela, R. Analysis of barbiturates by micro-high-performance liquid chromatography with post-column photochemical derivatization, *J. Chromatogra. A* 2000, *870*, 39-44.

12. Kyoko, I., Yoko, K., Hiroko, M., Toshiko, U. Determination of barbiturates in mouse tissue by high-performance liquid chromatography, *J. Chromatogra. A* 1981, *205*, 401-412.

13. White, P.C. Use of dual-wavelength UV detection in high-performance liquid chromatography for the identification of barbiturates, *J. Chromatogra. A* 1980, *200*, 271-276.

14. Hulshoff, A.H., Roseboom, J. R. Improved detectability of barbiturates in high-performance liquid chromatography by pre-column labelling and ultraviolet detection, *J. Chromatogra. A* 1979, *186*, 535-541.

15. Dilranjan, N., Pillai, S. D. Analysis of barbiturates by gas chromatography, *J. Chromatogra. A* 1981, *220*, 253-274.

16. Mats, G., Inga, P. Gas chromatographic determination of barbiturates by extractive alkylation and support coated open tubular column separation, *J. Chromatogra. A* 1977, *140*, 165-169.

17. International Conference on Harmonization (ICH), Q2(R1): Validation of analytical procedures: Text and Methodology, November 2005.

18. Reviewer Guidance: Validation of Chromatographic Methods, Food and Drug Administration (FDA), Center for Drug Evaluation and Research (CDER), November 1994.

19. Shabir, G.A. Validation of high-performance liquid chromatography methods for pharmaceutical analysis: Understanding the differences and similarities between validation requirements of the US Food and Drug

Administration, the US Pharmacopeia and the International Conference on Harmonization. *J. Chromatogr. A* 2003, *987*, 57-66.

25 METHOD DEVELOPMENT AND VALIDATION FOR THE GC-FID ASSAYOF P-CYMENE IN TEA TREE OIL FORMULATION

Finally in this research programme, the validation approach was evaluated as applied to another analytical technique. Here, GC was successfully used to develop a novel assay for p-cymene content in tea tree oil formulation. This presented a different analytical problem because of the very complex nature of this natural product, GC was the technique of choice for this method.

The use of essential oils in complementary medicine, particularly aromatherapy and also in the cosmetic and perfumery industry is becoming increasingly popular. The essential oil, which is distilled from the leaves of a tree *Melaleuca alternifolia*, commonly called tea tree is well known for its antimicrobial activity and has enjoyed increased medicinal uses in recent years. Oil of *Melaleuca alternifolia* (terpinen-4-ol type) is now clearly defined by the Draft ISO 4730. It sets a p-cymene content of not less than 0.5% and not more than 12%. Chemically p-cymene (**I**) is a 1-methyl-4-(1-methylethyl) benzene (Merck Index, 1996) occurs in a number of essential oils.

p-Cymene (I)

p-Cymene concentration can rise to levels approaching its upper limit. Two pathways are operating here (Figure 1), one involving hydrolysis of the pi bond at C-4 to produce terpinen-4-ol, the other involving oxidation of the p-menthane skeleton to its benzene analogue, p-cymene. The first pathway must involve water, naturally present in the oil through the steam distillation extraction process and possibly, trace volatile organic acids that may catalyse the reaction. The second pathway is well known in terpene chemistry. However, the various oxidation agent and catalysts that are involved in tea tree oil degradation required further investigation.

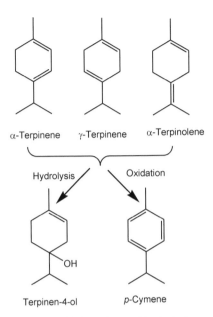

α-Terpinene   γ-Terpinene   α-Terpinolene

Hydrolysis   Oxidation

Terpinen-4-ol   p-Cymene

FIGURE 1 End products of hydrolysis and oxidation of constituents of tea tree oil.

1   STEP-BY-STEP NEW METHOD DEVELOPMENT

A similar systematic strategy was used in developing a new method for p-cymene as discussed in Chapter 5. The method development for the assay of p-cymene was based on its chemical properties. p-Cymene ($CH_3C_6H_4CH(CH_3)_2$) is a non-polar molecule and, therefore, a non-polar solvent hexane (C_6H_{14}) was used as the diluent. The medium polarity five percent Carbowax was used for separation. The GC-FID parameters used in the method development were based on the boiling point (177.10 °C) and the flash point (47 °C) of p-cymene. The injection port and detector temperature were set to

220 °C and oven temperature was set to 100 °C. The oven programme was isothermal with a run time of 10 min. The head pressure was set to ensure a hydrogen flow of 36 mL/min. The split was then adjusted to 6:01. The solvent, column and acquisition parameters were chosen to be a starting point for the method development. However, the separations produced using these parameters were excellent. The retention time of *p*-cymene was approximately 8.45 min with good peak shape and USP tailing was approximately 1.0. Additionally, preliminary precision and linearity studies performed during the development of the method showed that the 1.5 µL injection volume was reproducible and the peak response was significant at the analytical concentration chosen. Diluting the standard and sample in hexane gave solutions that could be injected directly (without further dilution, filtration or centrifugation). Chromatograms of the resulting solutions gave very good peak shapes (Figure 2 and 3) and co-elution of excipients was not observed on the same retention time as *p*-cymene.

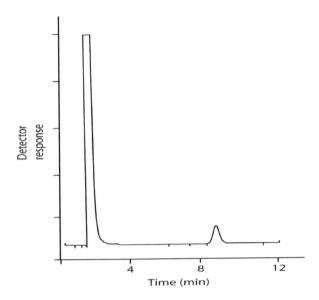

FIGURE 2 Separation of *p*-cymene reference standard. Conditions:
Phenomenex column packed 5% carbowax 20 m (6 x 0.25 mm) 80-100 mesh,
hydrogen gas, FID, injector and detector temperature 220 °C, oven
temperature 100 °C, flow rate 36 mL/min, injection volume 1.5 μL.

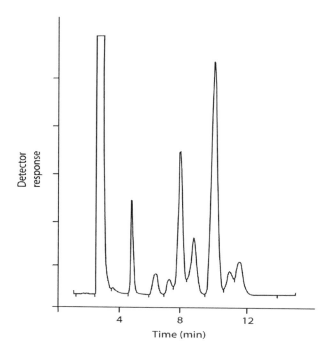

FIGURE 3 Separation of 100% tea tree oil sample. *p*-Cymene eluted at 8.46 min. Conditions: Phenomenex column packed 5% carbowax 20 m (6 x 0.25 mm) 80-100 mesh, hydrogen gas, FID, injector and detector temperature 220 °C, oven temperature 100 °C, flow rate 36 mL/min, injection volume 1.5 µL.

2   STEP-BY-STEP METHOD VALIDATION

For best practice, a similar step-by-step approach was adopted in validation of this assay method as used in HPLC assay in other chapters. Validation is a critical part of the development of a GC method. Although there is general agreement about what type of validation studies that should be done, there is great diversity in opinion as to how they are to be accomplished. The literature contains a variety of approaches to performing validation studies using gas chromatography (Robert & Eugene, 2004). This paper presents approaches that serve as a basis to perform validation studies for most GC methods in compliance with the pharmaceutical and biotechnology fields. The validation characteristics performed were linearity, range, specificity, accuracy, precision (repeatability and intermediate precision), limit of detection and limit of quantitation.

The linearity was studied in the concentration range 20-120 µg/mL. Six solutions were prepared corresponding to 20, 40, 60, 80, 100, and 120% of the nominal analytical concentration (100 µg/mL) and the following regression equation was found by plotting the peak area (y) versus the p-cymene concentration (x) expressed in µg/mL: $y = 354.05x + 0.218$ ($r^2 = 0.9995$). The analyte response is linear across 80-120% of the target p-cymene concentration.

The precision was examined by analysing six different samples by only one operator. The repeatability (within-run precision) was evaluated by only one analyst within one day, whereas reproducibility (between-run precision) was evaluated for three different days. The RSD values for within-run precision were 0.13% and for between-run precision was 0.66% (Table 1), therefore meeting the acceptance criteria for the chromatographic p-cymene assay.

TABLE 1 Within and between-run precision results for *p*-cymene in 100% tea tree oil formulation

Injection No.	Within-run precision		Mean between-run precision (n = 3 days)	
	Area (μV)	Retention time (min)	Area (μV)	Retention time (min)
1	29.2152	8.46	21.1505	8.45
2	29.0633	8.45	21.2953	8.45
3	29.6717	8.45	21.3440	8.45
4	28.3922	8.46	21.3291	8.46
5	28.5053	8.45	21.3453	8.45
6	28.2665	8.45	20.9958	8.45
Mean (6)	28.8523	8.45	21.2450	8.45
RSD (%)	0.13	0.06	0.66	0.05

The accuracy of the method was determined by fortifying sample with known amounts of the *p*-cymene reference substance at four concentration levels (0.5, 0.10, 0.05 and 0.01 mg/mL) and injected in triplicate. Mean recoveries for the samples analysed were found to be 95.3%.

The lower limit of detection for *p*-cymene was found to be (signal-to-noise = 3) 2.08 μg/mL. The LOQ values for *p*-cymene were found to be (signal to noise = 10) 10.39 μg/mL and RSD less than 2% for three replicate injections.

Assay interference (specificity) was investigated by injecting 12 month old stability solutions of tea tree oil. No interfering peaks were observed. Therefore, this method was specific for *p*-cymene.

3   STABILITY STUDY

The purpose of stability testing is to provide evidence on how the quality of a drug substance or drug product varies with time under the influence of a variety of environmental factors such as temperature, humidity and light, and enables storage conditions to be recommended and re-test and shelf lives to be established.

Tea tree oil is susceptible to oxidation and this is highlighted by the analytical data (Table 2) presented for two batches of tea tree oil. Samples of tea tree oil containing p-cymene content were packed in sealed dark brown glass bottles (10 mL) and stored at room temperature. The samples were withdrawn periodically (0, 1, 2, 3, 6, 9, 12 and 24 months) and the validated method was utilized successfully for analysing these stability samples. The data obtained from stability batches is evident that the assay is suitable for the quantitative analysis and stability testing of p-cymene content in the tea tree oil formulation.

TABLE 2 Stability data for p-cymene content in tea tree oil over 24 months

Batch #	Specification	Interval time in months							
		0	1	2	3	6	9	12	24
1	0.5-12%	2.2	2.8	2.7	2.4	3.1	3.0	3.5	10.1
2	0.5-12%	2.3	2.5	2.4	2.9	3.3	3.6	3.8	10.3

In this stability investigation, it could be seen that p-cymene concentration can rise to levels approaching its upper limit (maximum 12%). The key chemical indicator of oil treatment during processing and storage is p-cymene and elevated levels are usually an indication of poor storage conditions, old oil or harsh treatment during extraction.

Tea tree oil undergoes oxidation on storage via a number of different routes, some of which actually enhance oil, quality during the early stages. For long term storage for large quantities of neat oil stainless steel and aluminium, along with a nitrogen purge, provide the greatest stability (approximately two years). Dark glass with caps containing an impenetrable liner provides excellent long-term storage of small quantities. Stability of formulated products in glass is far superior to that in high-density polyethylene (HDPE) and polypropylene. However, laminates that incorporate aluminium or a fluorinated liner will enhance shelf life, although this may be dependent on the formulation matrix and tea tree oil concentration.

4   CONCLUSION

A new analytical method for the assay of p-cymene from a tea tree oil formulation was developed and validated using validated gas chromatography system. The validation study showed good linearity, (r^2 = 0.9995) sensitivity, accuracy and precession (RSD ≤ 0.66%). The proposed procedure was used in a quality control laboratory for analysis of formulations containing p-cymene products for final release.

This research has made a significant and coherent contribution to new knowledge including new validated GC method; new information on active p-cymene content; new information on tree tea oil product; a more detailed approach to method validation; a robust validated method and newly identified knowledge about product and shelf life extension.

REFERENCES

1. A. Penfold, 'Some Notes on the Essential oil of *Melaleuca* alternifolia', Australian J. Pharmacol. March (1937), 274-279.
2. G. Swords and G. L. K. Hunter, 'Composition of Australian Tea-Tree Oil (*Melaleuca* alternifolia)', J. Agric. Food Chem. 26 (1978) 734-739.
3. AT-2832, The Merck index, 12 (1996), 467-467.
4. Southwell I.A., Hayes A.J., Markham J., Leach D.N., The Search for Optimally Bioactive Australian Tea Tree Oil, Acta Hort. 344 (1993) 256-265.
5. Carson, C.F. and Riley, T.V., Antimicrobial Activity of the Major Components of the Essential Oil of *Melaleuca* alternifolia. J. App. Bacteriology, 78 (1995) 264-269.
6. Leach D.N., Wyllie S.G., Hall J.G. Kyratzis I., The Enantiomeric Composition of the Principle Components of the Oil of *Melaleuca* alternifolia, J. Agric. Food Chem. 41 (1993) 1627-1632.
7. Leach, D.N., Quality Aspects of Tea Tree Oil, Conference Proceedings "Tea Tree Oil – From Folklore to Fact" Australian Tea Tree Oil Export and Marketing Ltd., National Conference, Sydney, August 24-25 (1995).
8. ICH, Validation of Analytical Procedures. International Conference on Harmonisation, Geneva, Switzerland, March (1995).

DGS

PharmaTraining Ltd

Success through our experts

+44(0) 774 773 4950

gshabir@dgspharma.com

www.dgspharmatraining.com

DGS

PharmaTraining Ltd

DGS PharmaTraining Ltd is a UK-based global pharmaceutical training and consultancy services company that offers a diverse range of high quality training solutions for all levels of staff. We also offer an outstanding auditing service against all the major international regulatory and GMP standards including World Health Organisation (WHO), Medicines and Healthcare products Regulatory Agency (MHRA), U.S. Food and Drug Administration (FDA) and many more. Whatever your goals, we are prepared to help you as efficiently, expeditiously and economically as possible.

DGS PharmaTraining Ltd

Oxford, United Kingdom

www.dgspharmatraining.com

A huge selection of Pakistani/Indian readymade Shalwar Kameez and Unstitched material at very competitive prices.

Contact Details:

SobiaCollections, Oxford, United Kingdom
Email: sobiacollections@gmail.com